Volcanism and Fossil Biotas

Edited by

Martin G. Lockley
Department of Geology
University of Colorado at Denver
1200 Larimer Street
Denver, Colorado 80204

Alan Rice
Department of Physics
University of Colorado at Denver
1200 Larimer Street
Denver, Colorado 80204

© 1990 The Geological Society of America, Inc.
All rights reserved.

All materials subject to this copyright and included
in this volume may be photocopied for the noncommercial
purpose of scientific or educational advancement.

Copyright is not claimed on any material prepared
by government employees within the scope of their
employment.

Published by The Geological Society of America, Inc.
3300 Penrose Place, P.O. Box 9140, Boulder, Colorado 80301

Printed in U.S.A.

GSA Books Science Editor Richard A. Hoppin

Library of Congress Cataloging-in-Publication Data

Volcanism and fossil biotas / edited by Martin G. Lockley.
     p.   cm. — (Special paper ; Geological Society of America ;
244)
   Includes bibliographical references.
   ISBN 0-8137-2244-6
   1. Volcanism—Congresses. 2. Fossils—Congresses.
3. Paleoecology—Congresses. I. Lockley, M. G. II. Series:
Special papers (Geological Society of America) ; 244.
QE461.V633   1990
560—dc20   90-30476
                                                                                     CIP

Cover Art: Late Cretaceous duckbill dinosaurs overcome by
volcanic ash fall, by Doug Henderson. See Hooker (1987) and
Horner (1988) for details.

10 9 8 7 6 5 4 3 2

Contents

Preface v

How volcanism affects the biostratigraphic record 1
   Martin G. Lockley

An early terrestrial biota preserved by Visean vulcanicity
   in Scotland 13
   W.D.I. Rolfe, G. P. Durant, A. E. Fallick, A. J. Hall,
   D. J. Large, A. C. Scott, T. R. Smithson, and G. M. Walkden

Preservation, evolution, and extinction of plants in Lower
   Carboniferous volcanic sequences in Scotland 25
   Andrew C. Scott

The role of volcanism in K/T extinctions 39
   Alan Rice

Plant successions and interruptions in Miocene volcanic
   deposits, Pacific Northwest 57
   Ralph E. Taggart and Aureal T. Cross

Taphonomy and sedimentology of Arikaree (lower Miocene)
   fluvial, eolian, and lacustrine paleoenvironments, Nebraska
   and Wyoming; A paleobiota entombed in fine-grained
   volcaniclastic rocks 69
   Robert M. Hunt, Jr.

Evidence of flora and fauna in the gardens and cultivated
   land destroyed by Vesuvius in A.D. 79 113
   Wilhelmina F. Jashemski

Preface

Traditionally the study of volcanism and the study of fossil biotas are considered to be two very different subdisciplines in the broad realm of geology. Students may have learned in the classroom that fossils occur only in sedimentary rocks, that igneous deposits are essentially unfossiliferous. When considering the broad spectrum of different igneous rock types, particularly volcanogenic sediments, such a perception is entirely unfounded.

Fossils occur in a variety of volcanogenic deposits of all ages from Precambrian to Recent. As shown by the specific examples cited in this volume, volcanism can contribute directly, particularly through rapid burial, and indirectly, through diagenetic processes, to the preservation of both plant and animal remains. Moreover the special circumstances associated with volcanogenic processes often result in exceptional preservation of biotas, leading to fossil accumulations that preserve a better record than that found in many nonvolcanogenic deposits.

Volcanism also contributes to landscape modification through the formation of volcanic islands, the damming of rivers, and other alterations of subaerial and subaqueous topography. The biotic response to such abrupt changes in the physiochemical landscape leads to the development of new plant and animal communities. These biotic changes represent plant successions, and other biotic colonization, migration, and redistribution phenomena that are almost instantaneous from a geological perspective.

Volcanism may also cause long-term changes in climate that affect the biostratigraphic record. It is generally accepted that the Earth's atmosphere is of secondary origin, having been derived at least in part from degassing during the first billion years of Earth history. From this early point in time the atmosphere and biosphere have been delicately balanced. Some, like Gaian proponent James Lovelock, consider the biosphere a powerful driving force. Others claim that the physical environment exerts a greater measure of control. As pointed out by George Wald, we generally tend to think that the environment plays the tune to which organisms must dance. Even if in many instances organisms substantially modify the environment, there can be little doubt that many of the major extinction and turnover events in Earth history have been attributed to physical perturbations such as climate and sea-level changes, meteorite impacts, etc. As indicated herein, there is substantial evidence that volcanism has played a significant role in shaping the biostratigraphic record.

The contributions include studies that focus on specific faunas and floras from the Paleozoic, Mesozoic, Cenozoic, and Recent, and give more or less equal attention to paleobotany and paleozoology. They probably raise as many questions as they answer, some of them fundamental to the understanding of how the biostratigraphic record is shaped. They clearly point to the need to understand the sedimentologic, stratigraphic, taphonomic, and diagenetic relation between fossils and their entombing sediments.

THE BOULDER CONFERENCE

In May of 1987, at the request of the Paleontological Society (Rocky Mountain Section), we convened a special symposium at the Geological Society of America, Rocky Mountain Section Meeting in Boulder, Colorado. We were asked by the Paleontological Society to choose an appropriate theme that would be of broad appeal to earth scientists, and we chose "Volcanism and Fossil Biotas."

Ten abstracts were accepted and published in the program, eight papers were read during the meeting, and a field trip was organized to visit the famous volcanogenic Oligocene fossil beds at Florissant National Monument.

The Rocky Mountain region is rich in Mesozoic through Recent volcanogenic deposits, but particularly those of Cenozoic age. With the exception of some Cretaceous marine ashes, most of these deposits are terrestrial, including fluvio-lacustrine and flood-plain accumulations, and many are moderately to richly fossiliferous. Partly for this reason, the symposium was somewhat biased toward Cenozoic volcanism and biotas. However, in at least two papers, there was significant discussion of the role of volcanism in the terminal Cretaceous extinctions. This subject is receiving increasing attention as an alternative to the extraterrestrial–asteroid-impact hypothesis. These and other papers in the symposium attracted a large and lively audience and generated prolonged debates in the ample discussion time available.

It became clear that the notion of various connections between volcanism and fossil biotas is exciting to geologists. This is in part due to the dramatic circumstances surrounding rapid-burial mass-mortality sites like Pompeii, but also in part reflects interest in the far-reaching and longer-term effects that volcanism may have on the biostratigraphic record through modification of substrates, paleogeography, and climate, as well as the forcing of extinction events and evolutionary turnovers through climatic perturbations.

Many of the participants requested approval to present their papers as mini-reviews or overviews, incorporating some new data and interpretation. In view of the lack of precedents for presenting information on the subject, we thought it appropriate to encourage such overviews in an attempt to clarify the current state of knowledge.

COMPILATION OF THE VOLUME

Given the post-Paleozoic emphasis of the symposium contributions, we sought to balance the final volume by encouraging contributions dealing with Paleozoic sites. In successfully soliciting two such contributions, the volume is broader in scope than it would have been had only the symposium contributions been compiled. Although several of the original symposium participants were unable to submit manuscripts, such omissions are compensated for by similar or related contributions and/or discussion of their research. Moreover, the editors have encouraged the review perspective in all manuscripts and incorporated references to all the original symposium contributions in the appropriate places.

CONCLUSIONS

The chapters in this volume reflect some of our current knowledge of the complex relation between volcanism and fossil biotas. They are not meant to be definitive works on particular aspects of these relationships, so much potentially important detail is omitted. Often, as in the case of diagenetic studies, the information is lacking simply because enough work has not yet been done. The recent Penrose conference "Volcanic Influences on Terrestrial Sedimentation" reached similar conclusions regarding our understanding of volcaniclastic sediments. More work is needed in these new and rapidly evolving fields, with a need for improved consensus and standardization of approaches and terminology.

The result of the symposium and this compilation is a series of overviews that suggest ways to approach the study of volcanism and fossil biotas—a starting point for geologists and paleontologists who will often fine that the two subjects are inextricably linked.

Martin G. Lockley
Alan Rice

How volcanism affects the biostratigraphic record

Martin G. Lockley
Department of Geology, University of Colorado at Denver, 1200 Larimer Street, Denver, Colorado 80204

ABSTRACT

Volcanism significantly affects the biostratigraphic record by contributing to special preservation of fossil floras and faunas, and by modifying the paleoenvironments in which they developed and evolved.

The main local or near-source influences of ash and tephra falls, and downslope pyroclastic and debris flows, are rapid burial of biotas that are either in situ or transported only short distances. Large eruptions may cause similar rapid burial farther from source.

Volcanism also affects the development of local substrates and the configuration of lakes and drainages near source, and on a regional scale, causes the construction of volcanic islands. In all cases, new habitats are provided for biotas that may subsequently be represented in the biostratigraphic record.

On the local, regional, and global scale, volcanogenic emanations of CO_2, H_2S, SO_2, and other gaseous and particulate material are a significant cause of climatic perturbation leading to temperature fluctuations, and in some cases, disequilibrium in global biogeochemical cycles. This suggests that volcanism may have contributed more to the Cretaceous-Tertiary extinctions and other turnover/extinction episodes than is usually supposed.

INTRODUCTION

A superficial consideration of the relation between fossils and volcanic or volcanogenic rocks might suggest that there are few connections. For convenience, students might be told that fossils are essentially absent from igneous rocks, occurring "by implication" in non-igneous rocks of sedimentary origin. Closer examination of the distribution of fossil floras and faunas reveals that there is little reason to make such generalizations. It is true that fossils rarely occur in intrusive igneous rocks, except in xenoliths (Chronic and others, 1969); however, fossils are frequently found in extrusive igneous deposits, and in such settings may be extremely well preserved, thereby providing paleontological information superior to that available from typical nonvolcanogenic sedimentary deposits. Even where this is not the case, fossils from such rocks need to be recognized for the information they do provide.

A preliminary review of the distribution of fossils in volcanogenic deposits suggests a number of interesting relations that merit further investigation. Fossils occur in association with volcanogenic deposits throughout the Phanerozoic, but are generally more abundant and better studied in younger deposits. This is, in part, simply a function of the progressive diversification of biotas during the Phanerozoic (Sepkoski, 1981) and the lack of a terrestrial record in pre-Devonian strata. There is, however, no shortage of fossiliferous volcanogenic marine deposits in the lower Paleozoic, particularly where there was intense volcanic activity (Lockley, 1984).

A more fruitful line of investigation appears to be in understanding the relation between volcanogenic processes and the resultant fossil accumulations. As suggested by Lockley and Chesson (1987), the problem can be approached in a number of ways. Volcanogenic deposits can be classified according to their areal (paleogeographic) extent, ranging from local extrusions affecting a few individuals to regional or even global-scale ash falls and emanations that affect large communities and biotas. Examples range from the occurrence of an individual rhinoceras in lava (Chappel and others, 1951) to the mortality and burial of locally and regionally distributed biotas (Tuttle and others, 1987; Voohries, 1987; Taggert and Cross, 1987; Fritz, 1986). On a global

Lockley, M. G., 1990, How volcanism affects the biostratigraphic record, *in* Lockley, M. G., and Rice, A., eds., Volcanism and fossil biotas: Boulder, Colorado, Geological Society of America, Special Paper 244.

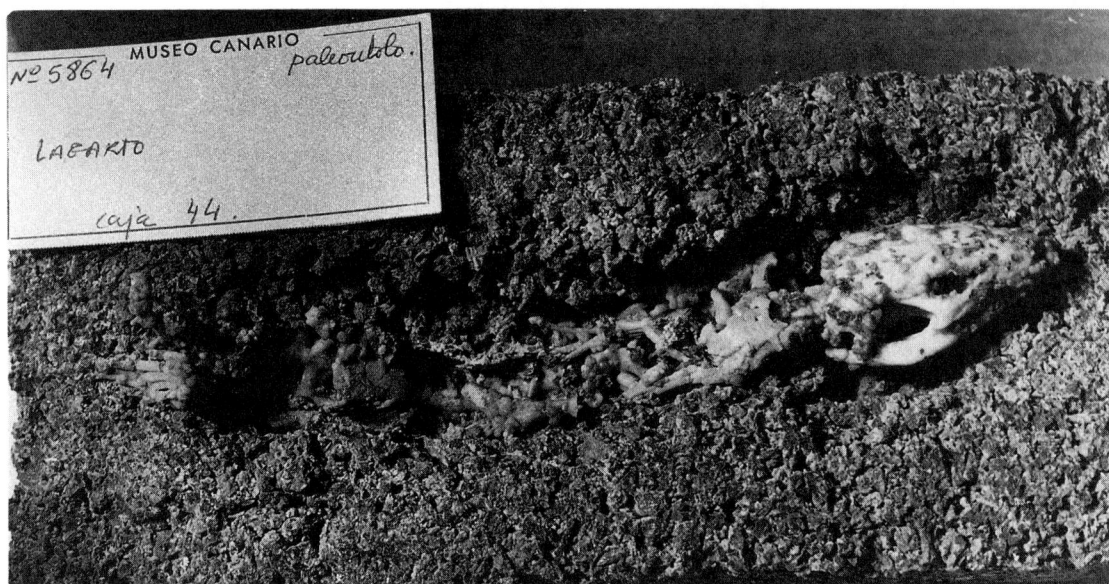

Figure 1. A lizard in Recent pyroclastic deposits (between 100,000 and 1,000,000 yr old). Specimen in Museo Canario, Las Palmas, Gran Canaria, Canary Islands. Photo by Karl Hirsch; magnification about 0.7 μm.

scale, volcanism clearly affects climate and evolution (Axelrod, 1981; Keith, 1982; McLean, 1985a, b, c), thereby significantly influencing the biostratigraphic record.

While these problems of scale are important, there is also much evidence to suggest that different volcanogenic processes affect the mode of preservation of biotas. Rapid burial by ash falls, pyroclastic surges, and lahars can lead to excellent preservation (Sigurdsson and others, 1985a; Gore, 1984; Naranjo and others, 1986). Similarly, volcanism may contribute directly to early diagenetic permineralization processes (Scott, 1987) or the generation of hot fluids, which result in increased mortality and preservation potential (Licht, 1987). Volcanism may also contribute to such phenomena as the lava damming of lakes and the construction of volcanic islands. Such lakes can become important depocenters for fossil biotas, while islands may act as endemic centers or land bridges for diverse biotas. It is also worth noting that volcanogenic soils and substrates are in themselves fertile ground, often permitting the establishment of distinctive plant and animal communities not known from other environments (Scott and Galtier, 1985; Scott this volume).

Other interesting associations between fossils and volcanic and volcanogenic rocks include Carboniferous and Cretaceous vent worm fossils associated with hydrothermal-spreading-center environments (Haymon and others, 1984; Oudin and Constantino, 1984), and fossil tracks and nests that owe their preservation to ash-fall episodes (Casamiquela, 1964; Burford, 1985; Hayward and others, 1982; Hay and Leakey, 1982).

FOSSILS ASSOCIATED WITH LAVA FLOWS

Because the temperature of molten lava is usually about 1,200°C, it is likely to burn up any organic material with which it comes into contact. However, the remains of large plants and animals may be preserved under certain circumstances. At Ginko State Park in the state of Washington, remains of Ginko logs and a Miocene rhinoceras *Diceratherium*, were found as pillow-lava moulds (Chappel and others, 1951), indicating rapid quenching of lavas as they surrounded organic remains in a swamp or pond environment. McKee (1979, p. 237) reported charred trees below lava flows in the Triassic Cave Sandstone sequence of southern Africa. A modern example of a lizard in lava has been reported by Hirsch (personal communication, 1986; Fig. 1) from a young volcanic sequence in the Canary Islands. As shown, the bones are preserved in a cavity, suggesting some similarities to the bone-bearing *Diceratherium* cavity. As discussed below, the "lava" may be more appropriately classified as a pyroclastic deposit.

FOSSILS ASSOCIATED WITH SUBAERIAL ASH AND TEPHRA FALLS

Explosive volcanism such as that witnessed at Mount St. Helens on May 20, 1987, or Nevado del Ruis, Columbia, in November 1985, produces Plinian eruption columns that eject large amounts of ash or tephra. The initial gas surge or basal ground surge may blast down trees and other objects for several kilometers around the vent. The orientation of fallen trees can provide useful directional indicators for paleogeographic reconstruction (Froggatt and others, 1981).

Subsequent fallout from Plinian columns is controlled by the magnitude of the convective upward thrust (Sparks and Wilson, 1976; Corey, 1985) and by prevailing winds, typically producing lobate ash-fall deposits that decrease in thickness in the downwind direction or distally, and that are often fossil rich near the source or wherever the accumulation is thick. In the examples cited above, isopach maps of the ashes indicated downwind accumulations in excess of 1 mm at distances of 500 and 50 km, respectively (Findley, 1981; Carey and Sigurdsson, 1982; Naranjo and others, 1986). The 1982 eruption of El Chicon produced similar accumulations up to 200 km downwind (Weintraub, 1982) and well-preserved floras in the vicinity (Burnham and Spicer, 1986).

Cook and others (1981) reported that the Mount St. Helens ashfall had widespread impact on agricultural vegetation, with accumulations of as much as 30 kg/m^2. In eastern Washington, crop losses were estimated at about $100 million in 1980. Ash severely affected some plants like alfalfa and caused high mortality in many insect populations. At Harper Island, 330 km northeast of Mount St. Helens, a colony of gull nests was buried by about 3.5 cm of ash by 0800 hours on May 19. A significant proportion of these nests were not excavated by the gulls, thus leaving eggs and nest structures under favorable conditions for fossilization (Hayward and others, 1982, 1989). Near the source of such eruptions, ash falls may accumulate to depths of several meters or more and cause the burial of buildings or other large structures (cf. Holmes, 1965, Fig. 216 for Sakurajima ashfall of 1914). Eruptions responsible for the formation of giant volcanic calderas (as much as 100 km in diameter) were orders of magnitude larger than any known in historic times; their associated ashfalls may be mapped on centimeter scale, not millimeter, over radii of thousands of kilometers (Francis, 1983).

In historic times, perhaps the most famous eruption, Vesuvius in A.D. 79, produced about 2.7 m of near-source pumice ash at Pompei (Sigurdsson and others, 1985a, b; Gore, 1984). Fallout initially contributed to the death and burial of biotas by ash (Jashemski, 1979), and was subsequently followed by even more lethal and destructive pyroclastic surges and flows. Similar processes associated with the eruption of Santorini evidently contributed to the destruction of Minoan civilization (Bond and Sparks, 1976). In these and other large historic eruptions, such as Mount Tambora (Stommel and Stommel, 1983), great volumes of ash fell in the marine environment as well as on land.

In the western hemisphere, Sheets (1981) has documented the periodic destruction of Mayan settlements by volcanic ash falls, notably a massive eruption around A.D. 260. Tilling and others (1984) have identified several Holocene eruptions of El Chichon and dated them at approximately A.D. 600, 1250, and 1700. Pottery found in soils between the pumice flow deposits associated with the two younger eruptions suggests dates between A.D. 800 and about 1200, showing that the archeological and radiometric dates are consistent.

Prehistoric, pre-Holocene eruptions, such as the one responsible for burying a remarkable Miocene vertebrate assemblage under a 2-m-thick ash unit in the Ash Hollow Formation of Nebraska, are responsible for the accumulation of thick ashes at considerable distances from the source (Voohries and Thomasson, 1979; Voohries, 1981). Similarly, the destruction and preservation of thousands of hadrosaurs (*Maiasaura peeblesorum*) in the Upper Cretaceous Two Medicine Formation of Montana has been attributed to a volcanic ash fall (Horner, 1988; Hooker, 1987). Such evidence leads to the conclusion that destruction of biotas and subsequent burial and accumulation is to a large degree directly correlated with the scale of volcanic activity and fallout.

FOSSILS IN SUBAERIAL VOLCANOGENIC GRAVITY-FLOW DEPOSITS

According to Leeder (1982, p. 76), gravity flows include grain, debris, liquified, and turbulent flows that "transport themselves with no help from the overlying stationary medium." Such categories more or less embrace a number of familiar volcanogenic flow processes including pyroclastic flows and surges, nuées ardentes, lahars, and debris flows. As demonstrated by Smith (1986), some of these categories (e.g., lahar) are poorly or inconsistently defined; others inadequately embrace distinctive categories of hitherto unrecognized volcaniclastic sediment. Unlike ash and tephra falls, which lead to vertical accretion of sediment, downslope gravity flows have a strong component of lateral movement and often transport biotic material before deposition and burial.

A distinction must be made between pyroclastic flows, surges, and nuées ardentes, which deposit hot ignimbrites and welded tuffs (about 500°C), and lahars and debris flows, which are much cooler (ambient temperatures), owing their existence to the mixing of volcaniclastic material with surface water and sediment. These two types of deposit result in different preservation and distribution of fossil biotas in and around volcanic centers.

As indicated above, pyroclastic surge and flow deposits often follow the initial Plinian ash and tephra fall episodes in an eruptive sequence. The heat, rapidity, and volume of the surges and flows may be responsible for the destruction of biotas on a scale exceeding that resulting from the ash fall. Sigurdsson and others (1985a) demonstrated that the majority of human remains discovered at Pompeii lay within pyroclastic surge deposits. They also demonstrated that some of the surges flowed at 100 m/sec, destroying buildings and transporting large blocks of wood and masonry. Rapid burial and preservation may result in good preservation if temperatures are not so high as to cause carbonization or complete destruction of organic material. Hirsch and Lopez-Jurado (1987) report eggs transported and preserved in Pliocene pyroclastic rocks from the Canary Islands (Fig. 2). The younger lizard-bearing deposits (Fig. 1) date between about 100,000 and 1,000,000 yr B.P. and also contain land snails (Lopez-Jurado, written communication, 1987). They are also partly pyroclastic in origin.

Lahars, debris flows, and hypoconcentrated flood flows (sensu Smith, 1986) are, like pyroclastic flows, highly destructive. Sediment flows resulting from the 1980 and 1982 eruptions of Mount St. Helens transported large trees for tens of kilometers downstream (Fritz and Harrison, 1985). Fritz and Harrison showed that log orientations are useful for reconstructing paleo-

Figure 2. Chelonian eggs in pyroclastic deposits from the Pliocene of Gran Canaria, Canary Islands. After Hirsch and Lopez-Jurado (1987) with permission; magnification 0.6×.

current directions and suggested that care must be taken in distinguishing between orientations arising from transportation and patterns arising from initial gas surge blasts (cf., Froggat and others, 1981).

Mudflows and lahars resulting in the melting of glacial ice by pyroclastic flow surges following the 1985 Nevado del Ruiz eruption in Columbia were confined to narrow ravines. They flowed fast (20 to 45 km/hr), tending to pulverize vegetation and organic matter (Fritz, 1986; Naranjo and others, 1986), and resulted in the deaths of more than 23,000 people.

Fritz (1986) has demonstrated that the Yellowstone fossil forest deposits of the Eocene Absaroka Volcanic Supergroup provide examples of fossil trees buried in situ, and others transported and oriented by volcanogenic mudflows. Thus, it is clear that, as with fossils in other sedimentary rocks, care must be taken to distinguish between in situ and transported remains in volcanogenic sediments.

Relatively few examples of pre-Cenozoic volcanogenic gravity-flow deposits have been documented. Recently, however, Fastovosky and others (1988) reported that Jurassic volcanogenic debris flows from the Huizachal Group of Mexico have yielded a diverse vertebrate fauna, including tritylodont, dinosaur, mammal, crocodylomorph, and pterosaur remains.

FOSSILS IN SUBAQUEOUS VOLCANOGENIC SEDIMENTS

In addition to pillow lavas and volcaniclastic sediments arising from subaqueous, mainly submarine eruptions, a great variety of subaqueous deposits are known that result from the fall or flow of various subaerial ashes and gravity flows into subaqueous environments. Many of these incorporate fossils.

In lacustrine environments, these fossils are of terrestrial origin, often including plants, aquatic invertebrates, fish, and insects (cf., MacGinitie, 1953; Licht, 1986, 1987). Subaerially generated volcaniclastic sediment may transport terrestrial biotic remains into the marine environment; however, most fossils in this depositional setting are marine, ranging from benthic to planktic forms.

As demonstrated by Carey and Sigurdsson (1980), airfall tephra may cover large areas of the ocean floor after settling through the water column. As with subaerial tephra deposits, the distribution is usually lobate or elliptical, controlled by prevailing winds. However, in marine environments, such accumulations may be more regular than on land because they are less susceptible to subsequent erosion or modification arising from the differential character of the land surface (vegetation, topography, etc.).

Fossiliferous volcanogenic sediments in marine sequences have given rise to a number of interesting stratigraphic names. For example, the Murchisoni Ash in the Ordovician of Wales is named after the zonal graptolite, and the Asaphus Ash after a trilobite (Williams and others, 1972). Although many of these so-called ashes are not true ashes, it is possible to find graptolites and a host of other marine organisms in volcanogenic deposits (Horne, 1976; Lockley, 1984; Neuman, 1972, 1976, 1984; Fig. 3).

A number of studies have noted a correlation between diverse marine faunas and volcaniclastic substrates. In the Ordovician of Wales a number of well-known fossiliferous units have been referred to as "calcareous ashes" or "calcareous tuffs" (Williams and others, 1972). The same terminology has been applied to younger successions, e.g., the Paleogene of New Zealand (Lee, 1986). Here brachiopod faunas associated with highly calcareous basaltic tuffs produced in submarine eruptions are two to three times more diverse than those associated with coeval mudstone, sandstone, and greensand substrates, and comparable with peak diversity levels recorded in the limestones (Lee, 1980, 1986; Fig. 4). Many of the brachiopod shells from these tuff beds are well preserved, with a large proportion of small individuals and articulated valves enclosing complete or partial internal void space (Lee, 1980, and personal observation). Such preservation indicates rapid burial. Although Lee (1980, 1986) infers some transportation, she concludes that most of the fauna is very nearly in situ. She also notes that there are few modern studies on the effects of volcanic activity on marine faunas. Fridriksson (1975) noted that fresh volcanic substrates around Surtsey were colonized within a few years. Often such "new" environments host diverse faunas, at least initially (Lockley, 1978).

High faunal diversity has also been noted in a number of volcanogenic mass-flow deposits. Hayward (1976, 1977a, b) has studied lower Miocene faunas from a number of biocoenoses and thanatocoenoses in the Waitakere Group of North Island, New Zealand. He demonstrated that transportation by mass-flow deposits led to the accumulation of fossil-rich volcorudites containing a diverse mixture of shallow to deep-water forms in a sequence of bathyal volcarenites and volcolutites. Similar conclusions were reached independently by Lockley (1984) in a study of a diverse Lower Ordovician fauna associated with a volcanic island complex at Builth Wells, Wales. Here, previously defined paleocommunities appear to be mixed in a volcanogenic debris flow (Suthren and Furnes, 1980). Despite the coarse-grained

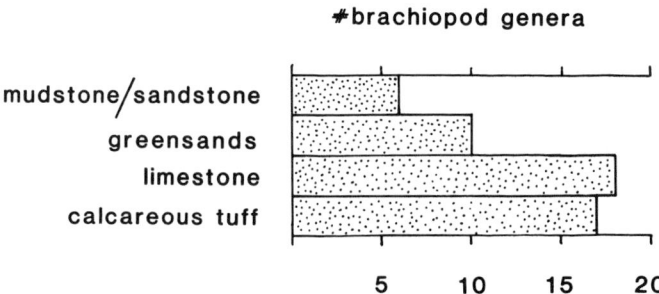

Figure 4. Diversity of Paleogene brachiopod faunas from New Zealand, arranged according to substrate. After Lee (1980, 1986).

character of the deposit, the fauna is remarkably well preserved, including very small brachiopods and largely unbroken bryozoan fronds and graptolites (Fig. 3).

Recent discoveries of fossils associated with ancient hydrothermal vents indicate new categories of biotas that may be preserved in the fossil record as a result of volcanic activity. Modern faunal assemblages associated with deep-sea spreading centers have only recently been discovered (Corliss and Ballard, 1977; Ballard and Grassle, 1979) and shown to represent unique biotic communities in which H_2S-ingesting chemoautotrophic bacteria form the base of the food chain. Haymon and others (1984) and Oudin and Constantino (1984) reported fossils associated with chimneys and mounds in the Cretaceous sulfide deposits of ophiolite complexes in Cyprus and Oman, respectively. Banks (1985) extended the known stratigraphic range of this type of deposit by reporting a fossil hydrothermal worm assemblage in lower Carboniferous ore-deposit chimneys from Ireland.

DIAGENETIC ASPECTS OF PRESERVATION OF BIOTIC REMAINS IN VOLCANOGENIC DEPOSITS

Little attention has been focused on the diagenesis of fossiliferous volcanogenic deposits, even though such research is essential for a better understanding of the contribution diagenesis makes to the preservation of biotas.

Rex and Scott (1987) and Scott (this volume) have demonstrated that Carboniferous plants in volcanogenic ashes can be preserved as calcareous permineralizations and as fusain. By contrast, chlorite permineralization is responsible for the preservation of Cretaceous leaves in Antarctic volcaniclastics from the Fossil Bluff Formation of Alexander Island (Jefferson, 1982). In many other instances, silicification of plant and animal remains is directly or indirectly attributable to the high silica content of volcaniclastic material (cf., Hooker, 1987; Durant and others, this volume). For example, Stokes (1945), who suggested that "volcanic activity may have contributed to the death of the dinosaurs" at the Cleveland Lloyd Quarry (Morrison Formation of Utah), also inferred that "the siliceous nature of the bentonitic matrix in which the dinosaurs were buried is probably responsible for their preservation." Stokes's observations are similar to those made by

Figure 3. *Didymograptus murchisoni* from an upper Llanvirnian volcaniclastic debris-flow deposit, near Builth Wells, Wales. After Lockley (1984); magnification 10×.

Hooker (1987) and Horner (1988) with respect to the silicification of bones at the hadrosaur bone bed in Montana.

In younger deposits, such as those at Pompeii, it has been noted that the preservation of biotas varies according to local diagenetic conditions, such as ground moisture and the physiochemical characteristics of the deposit (Jashemski, 1979; Sigurdsson and others, 1985a; Sigurdsson personal communication, 1987). Hayward and others (1989) noted that gull eggs became firmly cemented in place when the Mount St. Helens ash in which they were buried was moistened. Hirsch (personal communication, 1989) indicates that the preservation of egg shell in such situations is controlled by the chemical composition of the entombing ash.

VOLCANISM AND PALEOBIOGEOGRAPHY

Plate-tectonic theory has made it clear that volcanism plays a major role in the changing configuration of continents and ocean basins through geologic time. Volcanism at mid-ocean ridge spreading centers results in the construction of new ocean crust, and the rate at which this volcanism proceeds may determine the elevation of the mid-ocean ridge and, in turn, affect sea level and climate (Berner and others, 1983). Where plates collide, volcanism is also ubiquitous, either in the form of oceanic island arcs or in association with mobile belts along continental margins. Almost anywhere that volcanism occurs the local paleogeography will be subject to modification, in turn affecting the paleobiogeography. Although it could be argued that the entire biostratigraphic record is directly attributable to global-scale constructive and destructive volcanogenic processes that constitute the plate-tectonic phenomena, such a discussion is daunting in scope. This discussion, therefore, is confined to specific local examples from the terrestrial and marine realm.

Terrestrial volcanism changes the configuration of the landscape through a number of processes, including accretion of lava and volcanogenic sediment, local erosion, local damming of drainages to form lakes, destruction of vegetation, and the formation of new substrates (soils) for colonization. All these volcanogenically induced paleogeographic changes are potentially recognizable in the stratigraphic and biostratigraphic record. Several of these topics are dealt with elsewhere (Scott and Galtier, 1985; Taggert and Cross, 1987). The following discussion deals only briefly with the subject of new substrates and lake formation.

New volcanogenic substrates provide new and often fertile ground for colonization by plant communities. In some cases, specially adapted forms, such as the nitrogen-fixing tree *Myrica faya*, invade new volcanic substrates, leading to the development of ecosystems that are entirely different from those that inhabited the underlying substrates (Vitousek and others, 1987). It has been noted (Spicer and others, 1985) that the first colonizers are often ferns. This observation has considerable implications for the biostratigraphic record. Floras and palynofloras are often likely to represent seres or successional stages in plant-community development, and not well-developed climax communities (Taggart and Cross, this volume). Such evidence also has an important bearing on the interpretation of the Cretaceous-Tertiary boundary fern-spore spike discussed below.

Damming of local drainages by lava or other volcanogenic sediment can create new lakes in subaerial environments (Buesch, 1988). Such lakes are focal points for the development of new biotic communities and, because they are depocenters, are likely to be conducive to the accumulation of representative biotic remains in the fossil record. Following the 1980 eruption of Mount St. Helens, Spirit Lake became substantially enlarged, and a small lake developed on Coldwater Creek as a result of blockage by debris flows (Findley, 1981, p. 21). The creation of new subaqueous environments in terrestrial settings is analogous to the formation of an island in an aquatic environment. The new biotas that become established show abrupt ecological change relative to the surrounding environments, and often exhibit unusual endemic characteristics.

Although lake deposits are common features of many terrestrial successions, few can be directly attributed to volcanic activity. Possibly the best example in the stratigraphic record is the Oligocene site at Florissant, Colorado, where lake formation is attributed to lava damming (MacGinitie, 1953). The Florissant flora and insect-rich fauna are world famous, and owe their preservation directly to the formation of lakes and the ongoing Oligocene volcanic activity in the area (McLeroy and Anderson, 1966; Licht 1986, 1987). Another example is the 8,000-yr-old Ampasambazimba lake basin site in Madagascar, also the result of lava damming (Burney and MacPhee, 1988).

In the marine environment, the formation of volcanic islands results in local development of a terrestrial habitat that can become quickly colonized (Fridriksson, 1975). The island will also be fringed by shallow-water environments, which soon develop characteristic facies faunas. Darwin (1859) showed that islands are ideal laboratories for natural experiments in evolution and that unique endemic forms can evolve over short periods of geologic time. However, terrestrial biotas are vulnerable to erosion, particularly in island settings, and it may be safely assumed that the fossil record of such biotas is sparse. One possible exception was suggested by Nopcsa (1923, 1934) for the unusual Late Cretaceous fauna from Transylvania, which he claimed had evolved in isolation on an island that he christened "Sienbenburgen." Although his suggestion needs further critical evaluation (Weishampel and Reif, 1984), it is known that this part of the Transylvanian region was, in fact, part of a volcanic island arc during the Late Cretaceous (Burchfiel and Bleahu, 1976).

Volcanic islands also provide habitats for peri-insular marine faunas, which have sometimes been described as "curious" (Horne, 1976) or "peculiar" (Bruton and Harper, 1981a, b). The unusual nature of these faunas is sometimes ascribed to a high degree of endemism (Neuman, 1972, 1976, 1984). Although endemism is to be expected, it can only be measured relative to other coeval faunas, which may be incompletely represented in the fossil record. There is also the problem of differential preservation referred to above (Lockley, 1984). Nevertheless, the marine and, to a lesser extent, terrestrial environments associated with volcanic islands are represented in the biostratigraphic record and have been cited as potential land bridges for the migration of faunal stocks (Hunt, 1987), as well as centers for the evolution of endemic forms. A good example of this is provided by Auffenberg (1981) in his discussion of the Komodo Dragon.

VOLCANISM AND CLIMATE

Much has been written about the influence of volcanism on climate, and the implications of such influences for evolution and extinction (Axelrod, 1981; Keith, 1982; Sigurdsson, 1982; Sigurdsson and Carey 1988; McLean, 1985a, b, c). Climates may be viewed from local, regional, or global perspectives and in terms of short- and long-term changes. Among the many important factors that contribute to large- and small-scale changes are fluctuations in temperature and changes in the gaseous and particulate composition of the atmosphere. Such changes may be picked up in the geologic record, as demonstrated by Tilling and others (1984) in their correlation of Holocene eruptions of El Chichon with acidity peaks in Greenland ice cores.

It is obvious that volcanic eruptions cause dramatic changes in the local climate by elevating temperatures to lethal levels and discharging all manner of particulate and gaseous material. There are, however, other forms of volcanogenic emanation, not directly associated with eruptions, that can severely affect local climate.

The recent lethal gas bursts from Lake Nyos in Cameroon, West Africa, caused the deaths of more than 1,700 people and 8,000 animals by CO_2 asphyxiation (Sigurdsson, 1987; Tuttle and others, 1987). It is evident from the study of this lake and others—such as nearby Lake Monoun which caused similar fatalities in 1984—that the CO_2 is derived from magmatic sources. Evidently, mantle-derived CO_2 accumulated and seeped into the anoxic bottom waters (hypolimnion) of the young crater lake. Sudden degassing was probably the result of an upset of the lake's density stratification by a local landslide (Sigurdsson, 1987). Once released, an estimated 1.2 km^3 of CO_2 flowed over the crater rim and into adjacent valleys, claiming victims as far as 20 km from the lake, before dissipating. Such natural disasters cause significant mortality, which may exceed that resulting from many historic volcanic eruptions. However, because the gas release is not accompanied by tephra fall or sediment-rich gravity flows, the potential for burial of deceased biotas is low and the likelihood of fossilization significantly reduced. However, it should not be assumed that such mortality would not show up in the fossil record. Subsequent reworking of mass mortality victims by normal sedimentation processes cannot be ruled out (cf., Hooker, 1987).

Larger-scale volcanic influences on climate include regional or global-scale temperature changes and other phenomena that can be attributed directly to specific volcanic events (Sear and others, 1987). Examples include the climatic perturbations attributed to the Laki eruption of 1783 (Sigurdsson, 1982) and the Tambora eruption of 1815 (Stommel and Stommel, 1983). Longer-term climatic changes may be attributed to major episodes of volcanism. Examples include episodes of Tertiary explosive volcanism, cited by Axelrod (1981) as the cause of sharply lowered temperature influencing plant and mammal evolution, and the suggestion that major episodes of flood-basalt extrusion have contributed to global climatic change and biotic evolution throughout much of the Phanerozoic (Keith, 1982; Morgan, 1986; McLean, 1981a, 1985a, b, c).

Morgan (1986) has drawn attention to the possibility of a causal relationship between flood-basalt episodes and extinctions during all Phanerozoic eras. Others, like McLean (1981a, 1985a, b, c), have focused attention on the terminal Mesozoic extinctions. They suggest that the high-volume, short-duration discharge of the Deccan traps flood basalts elevated atmospheric CO_2 levels, causing a severe greenhouse effect (McLean, 1978). McLean's hypotheses are particularly interesting in the light of the heated debate surrounding the K/T extinctions. Although a thorough review of this subject is beyond the scope of this chapter, it is appropriate to consider the extent to which volcanism may have contributed. For a recent thorough review see McCartney and Loper (1989).

It is known that historical eruptions have caused severe climatic disturbances leading to local- and regional-scale mortality of biotas as a result of phenomena other than inundation and burial by volcanogenic material. The Laki fissure eruption of 1783 produced the largest lava flow (565 km^2) known in historic time, but very little ash (Thorarinsson, 1969). Although the lava ran over settlements, the main damage was from "toxic volcanic gases, which caused the catastrophic haze famine of Iceland" (Sigurdsson, 1982, p. 601). An estimated 1.3 to 1.6×10^7 tons of SO_2 was released, resulting in stunting of pasture and the resultant death of 50 percent of the cattle, 79 percent of the sheep, and 76 percent of the horses. The human population was decimated by famine; more than 10,000 people, 24 percent of the total population, perished (Thorarinsson, 1969; Gledhill, 1986; Wood, 1984; Rice, this volume).

It is evident that the climatic effects of this eruption were widespread. In 1784, Benjamin Franklin noted that during the previous summer "there existed a constant fog over all of Europe and a part of North America." He also suggested that the cause might be related to "the vast quantity of smoke, long continuing to issue during the summer from Hecla in Iceland." According to Sigurdsson (1982), who cited these quotations, such observations represent the first recognition of the effects of volcanism on climate. However Franklin was not the only scientist of the day to record the strange weather conditions. According to Lockley (1976), Wood (1984), and Gledhill (1986), Gilbert White (1936) also recorded the "haze or smokey fog" and its effect on the 1783 summer weather in Britain. "At the time there was no true explanation of the phenomenon, and country people regarded the sun's lurid aspect, short of its beams, especially at sunrise and sunset, with superstitious forebodings" (Lockley

1976, p. 85). Sigurdsson (1982) summarized the paleoclimatic evidence obtained from historical temperature records and Greenland ice cores. This evidence indicates unusually cold mean temperatures, the lowest since records began, in the winter of 1783–1784. Similarly, acidity levels in the 1783 ice layer are higher than any recorded in the last 1,000 years. They are attributable to an estimated 100×10^9 kg of H_2SO_4 fallout derived from the atmospheric conversion of volcanogenic SO_2 and H_2S (Hammer, 1980).

The 1815 eruption of Mount Tambora also resulted in severe deterioration of the weather. Stommel and Stommel (1983) characterized 1816 as "a year without summer" and demonstrated well-documented crop failures and agricultural havoc in New England and Europe.

These two examples indicate that both fissure and explosive eruptions can cause severe climatic perturbations leading to global-scale impact on biotas. Although the causes (ash and tephra versus gases and aerosols) may be different, the fact that relatively small and short-lived historic eruptions can have such far-reaching effects, strongly suggests that larger prehistoric eruptions (Francis, 1983) must have had even more severe and long-term influences on climate.

The Deccan traps flood-basalt episode produced an estimated 2.6×10^6 km^2 of lava (McLean, 1985a, b). This is approximately 4,600 times the area covered by the Laki eruption lavas. The eruptions, which continued for about 0.5 to 1.5 m.y., produced an estimated 5×10^7 moles of CO_2. This probably represented a 10 to 25 percent increase in the rate of mantle production of CO_2. Oelofsen (1978) has also suggested that volcanism could contribute as much or more CO_2 to the atmosphere as is currently produced by the burning of fossil fuels, which has raised atmospheric levels by 10 to 12 percent since the turn of the century.

In 1978, McLean proposed a trans–K/T carbon-cycle perturbation characterized by CO_2-induced greenhouse conditions. Oelofsen (1978), reasoning along similar lines, spoke of an atmospheric CO_2/O_2 imbalance caused by kimberlite volcanism, basalt flows, and a reduction in O_2 production by marine phytoplankton. In 1981a, McLean specifically referred to a link between the CO_2 greenhouse conditions and Deccan traps volcanism. For some time the ideas presented by these authors were overshadowed by the announcement of Alvarez and others (1980) that the K/T boundary was marked by an iridium-rich clay, which they attributed to the impact of a large meteorite. De Laubenfels (1956) proposed a similar cause long before iridium had been identified at or near the boundary. This extraterrestrial hypothesis, implying sudden extinctions, has gained much attention and popularity in the scientific and popular press, and stimulated much research on the K/T boundary; however, there are a number of lines of evidence that suggest terrestrial causes and less abrupt extinctions (Rice, this volume). In 1983, Zoller and others reported "iridium" enrichment in airborne particles from Kilauea volcano," thus demonstrating that iridium does not have to have an extraterrestrial origin. The following year the meteorite scenario gathered additional support when Bohor and others (1986) reported shocked minerals at the K/T boundary, inferring decisive evidence for an impact. However, Rice (1987a and b) has demonstrated that explosive volcanism can form shocked minerals, a conclusion supported by the discovery of such minerals in Recent volcanic ash (Carter and others, 1986).

The conclusions of many workers who infer a gradual-extinction model rather than the sudden-death models implied by the impact scenario are also worth considering. If gradual extinction can be demonstrated (cf. Archibald and Clemens, 1982; Sloan and others, 1986; Ward and others, 1986; Whaley, 1987), the sudden-impact scenario must be seriously reevaluated. Although these authors do not relate the gradual-extinction evidence directly to volcanism, they generally attribute it to some form of climatic change. Others, like Keith (1982), Donovan (1987), and Sahni (1988), infer that volcanism was directly responsible. Keith (1982) concluded that the extinctions were protracted and related to the long-term biogeochemical-cycle fluctuations recorded in the isotopic record. Sahni (1988), who also regards the extinctions as gradual and has first-hand experience of the Deccan traps successions, generally favors a volcanic explanation and concurs with McLean (1978) that the Deccan traps degassing was the causative factor. He summarizes the current extinction debate by suggesting that it revolves around two hypotheses: the DT (Deccan traps) and the ET (extraterrestrial) scenarios. McCartney and Loper (1989) suggest that deep-rooted mantle plume/kimberlite volcanism is emerging as a paradigm that rivals the impact scenario.

It is worth mentioning one example of sudden biotic (floral) change at the iridium layer recorded in the Raton Basin, New Mexico (Orth and others, 1981), which might appear to be evidence of abrupt change. Angiosperm pollen declines rapidly, giving way to a relatively large increase in fern pollen. Such evidence can appear to be quite consistent with a catastrophic event such as an asteroid impact, but is also consistent with the anticipated pattern of recolonization seen on new volcanic substrates (Spicer and others, 1985). In several of his papers, McLean has pointed out that abrupt and simultaneous terminations of taxa are not always the result of a single catastrophic event. They may simply indicate a stratigraphic hiatus or other perturbation of the stratigraphic record.

CO$_2$, DINOSAUR EGGS, AND REPRODUCTION

The above-mentioned volcanic, CO_2-induced greenhouse scenario for the terminal Cretaceous is appealing because it provides plausible explanations for events in both the terrestrial and marine realms. Elevated atmospheric CO_2 levels are likely to have caused reproductive stress among the Dinosauria and other animal groups (McLean, 1978, 1986). Eggs must exhale CO_2 during normal respiration, and gas conductance is important for embryonic survival (Ferguson, 1985). This process will be inhibited by increased partial pressure of CO_2 in the atmosphere or nest mound (Seymour, 1979; Seymour and Ackerman, 1980). Large eggs and those of endotherms will suffer most because of

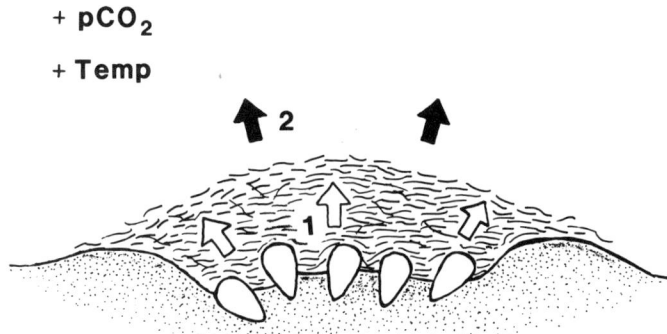

Figure 5. Model showing the effect of greenhouse conditions on respiration in dinosaur eggs and nests. CO_2 flow from egg to nest (1) and from nest to atmosphere (2) is inhibited by elevated pCO_2.

their surface-to-volume ratio and higher embryonic respiratory requirements (Oelofsen, 1978). Dinosaur eggs may have suffered from what could be called a "Russian doll effect" (eggs within clutches, within nests, within a high-CO_2 atmosphere). Elevated atmospheric CO_2 levels would inhibit diffusion of CO_2 from nest mounds into the atmosphere, in turn inhibiting diffusion of CO_2 from the egg to the nest (Fig. 5). As suggested by Oelofson (1978), ectotherms and endothermic layers of smaller uncovered eggs (i.e., many birds) would have been less vulnerable to elevated CO_2 levels.

According to Seymour and Ackerman (1980) and Williams and others (1984), Upper Cretaceous dinosaur eggs show pore adaptations to nest atmospheres that are high in CO_2 and low in O_2. Such adaptations resemble those seen in turtles, crocodiles, and megapode birds, and are thought to be related to underground nesting where water-vapor saturation levels are high. In the case of extinct dinosaurs, there is no proof that they habitually buried their eggs. Pore adaptations reflecting high CO_2 levels may reflect higher atmospheric concentrations of this gas. If the eggs were buried, they would be vulnerable to suffocation. If they were at or near the surface, they would be vulnerable to lowered humidity, which would cause excessive dehydration.

Assuming that the eggs were buried in environments similar to those created by modern birds and reptiles that build underground nests, gas conductance would have been affected by changes in pCO_2. If atmospheric CO_2 levels were raised (cf., McLean, 1985a, b) then CO_2 diffusion from eggs and nests would be retarded. All other factors being equal, a doubling of atmospheric CO_2 would cut the diffusion rate in half. Such changes could lead to excessive accumulations of CO_2, resulting in changes in temperature, humidity, and pH levels in nest environments.

It has been noted that the majority of known dinosaur eggs are Late Cretaceous in age. Two major sites—the Aix region of the Pyrennes (Erben and others, 1979), and peninsular India (Jain, 1989; Sahni, 1989)—are associated with the K/T boundary. If hatching proved to be a serious problem, for whatever reason, then we might expect to see a concentration of unhatched clutches. Based on estimates from hatcheries of 20 km^2 or more (Jain, 1989), this could be the case. It is also of considerable interest that these widespread hatcheries are associated with infra- and intra-trappean sites, closely associated with the Deccan lavas. As Sahni (1989) points out, dinosaurs were abundant in the region when the eruptions began. There is currently no record of them in beds overlying the Deccan traps.

It has recently become clear that critical fluctuations in the physio-chemical environment, notably temperature, may be crucial for embryonic survival. For modern reptiles like *Alligator mississippiensis,* temperature plays a critical role in sex determination (Ferguson and Joanen, 1982, 1983). Ideas about temperature-induced sterility and evolution are not new (Cowles, 1945), but they are currently receiving much scientific attention (McLean, 1981b, 1987). Paladino and others (1989) suggest that temperature-dependent sex determination may have played a crucial role in dinosaur extinction. It at least appears logical to assume that dinosaur reproduction would be affected by temperature to some degree, in much the same way as it affects other vertebrates.

THE MARINE REALM

The greenhouse scenario also fits well with biostratigraphic and isotopic evidence from the marine realm, indicating trans–K/T warming. It need not be analyzed here in detail (see references by McLean, and Rice, this volume, for further discussion and references). The extinction of calcareous marine plankton and the onset of "Strangelove" ocean conditions is well documented (Hsü and McKenzie, 1985; McLean, 1985a, b, c), and convincingly interpreted to be the result of CO_2 accumulation in the mixed layer and failure of the Williams Riley pump. Deposition of trace-element-enriched clays is a natural consequence of reduced carbonate deposition and perturbation of the carbon cycle (Keith, 1982). Much of the evidence presented by Ward and others (1986), and marine K/T-boundary biostratigraphers, pertains to gradual, climatically induced changes occurring over time periods on the order of 10^5 to 10^6 yr. Keith (1982) and Donovan (1987) cite volcanism as being directly and indirectly responsible.

CONCLUSIONS

(1) Volcanism has a widespread influence on the biostratigraphic record, and its effects may be seen on a local, regional, or global scale.

(2) Local volcanic centers may give rise to rapidly deposited volcanogenic sequences, which entomb biotas through rapid mortality and burial. Larger volcanic centers may bury biotas far from their sources.

(3) Diagenesis of volcanogenic deposits contributes to diverse modes of preservation, and is a field in need of further study.

(4) Volcanism may contribute to significant changes in the configuration of marine and terrestrial paleoenvironments, as

with the formation of volcanic islands and lakes. Such centers act as focal points for the development of local, sometimes endemic faunas, which may show up in the biostratigraphic record.

(5) Volcanism affects climates and biogeochemical cycles on a local, regional, and global scale, and over short and long time spans.

(6) Volcanically induced changes in climate and the equilibrium of biogeochemical cycles contribute to major evolution and extinction episodes in Earth history. Such changes may affect organisms in the areas of reproduction, embryogenesis, and sex determination.

ACKNOWLEDGMENTS

I was prompted to convene the Volcanism and Fossil Biotas symposium at the Rocky Mountain Section meeting of the Geological Society America at Boulder, Colorado, in May 1987, by my fellow officers Ernest Gilmore and Richard Moyle of the Rocky Mountain Section of the Paleontological Society. I thank them and my co-convenor, Alan Rice, for help with the organization. Robert Chesson also helped with the convention and library research for this chapter. I am indebted to Dewey McLean, Karl Hirsch, Haldar Sigurdsson, and Alan Rice for useful discussions, and also to Dr. Lopez Jurado for written information and permission to publish Figure 1.

REFERENCES CITED

Alvarez, L. W., Alvarez, W., Asaro, F., and Michel, H. V., 1980, Extraterrestrial cause for the Cretaceous–Tertiary extinction: Science, v. 208, p. 1095.

Archibald, J. D., and Clemens, W. A., 1982, The Late Cretaceous extinctions: American Scientist, v. 70, p. 377–385.

Auffenberg, W., 1981, The behavioral ecology of the Komodo Monitor: Gainesville, University of Florida Press, 406 p.

Axelrod, D. I., 1981, Role of volcanism in climate and evolution: Geologic Society of America Special Paper 185, 59 p.

Ballard, R. D., and Grassle, J. F., 1979, Return to oases of the deep: National Geographic, v. 156, p. 689–703.

Banks, D. A., 1985, A fossil hydrothermal worm assemblage from the Tynagh lead-zinc deposit in Ireland: Nature, v. 313, p. 128–131.

Berner, R. A., Lasaga, A. C., and Garrels, R. M., 1983, The carbonate-silicate geochemical cycle and its affect on atmospheric carbon dioxide over the past 100 million years: American Journal of Science, v. 283, p. 641–683.

Bohor, B. F., Foord, E. E., Modreski, P. J., and Triplehorn, D. M., 1986, Mineralogical evidence for an impact event at the Cretaceous–Tertiary boundary: Science, v. 244, p. 867–869.

Bond, A., and Sparks, R.S.J., 1976, The Minoan eruption of Santorini, Greece: Journal of the Geological Society of London, v. 132, p. 1–16.

Bruton, D. L., and Harper, D.A.T., 1981a, Ordovician volcanic islands and their faunas in the Appalachian–Caledonide orogen [abs.]: Terra Cognito (Uppsala Caledonide Symposium), v. 1, p. 37.

—— , 1981b, Brachiopods and trilobites of the early Ordovician serpentine Otta Conglomerate, south central Norway: Norsk Geologisk Tidsskrift, v. 61, p. 153–181.

Buesch, D. C., 1988, Changes in depositional environments resulting from emplacement of large-volume pyroclastic flows: Geological Society of America Abstracts with Program, v. 20, p. 178.

Burchfiel, B. C., and Bleahu, M., 1976, Geology of Rumania: Geological Society of America Special Paper 158, 82 p.

Burford, A. E., 1985, Reptilian markings on the upper Mowry Shale, Emigrant Gap area, Natrona County, Wyoming: Wyoming Geological Association 36th Annual Field Conference Guidebook, p. 157–158.

Burney, D. A., and MacPhee, R.D.E., 1988, Mysterious island; What killed Madagascar's large native animals?: Natural History, v. 97, p. 46–55.

Burnham, R. J., and Spicer, R. A., 1986, Forest litter preserved by volcanic activity at El Chichon, Mexico; A potentially accurate record of pre-eruption vegetation: Palaios, v. 1, p. 158–161.

Carey, S., and Sigurdsson, H., 1980, The Roseau Ash; Deep sea tephra deposits from a major eruption on Dominica, Lesser Antilles arc: Journal of Volcanology and Geothermal Resources, v. 7, p. 67–86.

—— , 1982, Transport and deposition of distal tephra from the May 18, 1980, eruption of Mount St. Helens: Journal of Geophysical Research, v. 87, p. 7061–7062.

—— , 1985, The May 18, 1980, eruption of Mount St. Helens; Part 2, Modeling of dynamics of the Plinian phase: Journal of Geophysical Research, v. 90, p. 2929–2958.

Carter, N. L., Officer, C. B., Chesner, C. A., and Rose, W. I., 1986, Dynamic deformation of volcanic ejecta from the Toba caldera; Possible relevance to Cretaceous/Tertiary boundary phenomena: Geology, v. 14, p. 380–383.

Casamiquela, R. M., 1964, Estudios ichnologicos: Buenos Aires, Colegio Industrial Pio IX, 229 p.

Chappel, W. M., Durham, J. W., and Savage, D. E., 1951, Mold of a Rhinoceros in basalt, Lower Grand Coulee, Washington: Geological Society of America Bulletin, v. 62, p. 907–918.

Chronic, J., McCallum, M. E., Ferris, C. S., Jr., and Eggler, D. H., 1969, Lower Paleozoic rocks in diatremes, southern Wyoming and northern Colorado: Geological Society of America Bulletin, v. 80, p. 149–156.

Cook, R. J., Barron, J. C., Papendick, G. J., and Williams, G. J., III, 1981, Impact on agriculture of the Mount St. Helens eruption: Science, v. 211, p. 16–22.

Corliss, J. B., and Ballard, R. D., 1977, Oases of life in the cold abyss: National Geographic, v. 152, p. 441–453.

Cowles, R. B., 1945, Temperature induced sterility and evolution: Science, v. 101, p. 221–222.

Cross, A. T., and Taggert, R. E., 1987, Sources and rates of plant accumulation in deposits from recent eruptions of Mount St. Helens: Geological Society of America Abstracts with Programs, v. 19, p. 268.

Darwin, C., 1859, The origin of species by means of natural selection or the preservation of favored races in the struggle for life: London, Murray.

De Laubenfels, M. W., 1956, Dinosaur extinction; One more hypothesis: Journal of Paleontology, v. 30, p. 207–212.

Donovan, S. K., 1987, How sudden is sudden?: Nature, v. 328, p. 109.

Erben, H. K., Hoefs, J., and Wedepohl, K. H., 1979, Paleobiological and isotope studies of eggshells from a declining dinosaur species: Paleobiology, v. 5, p. 380–414.

Fastovsky, D. E., Hermes, O. D., Clark, J. M., and Hopson, J. A., 1988, Volcanos, debris flows, and Mesozoic mammals; Huizachal Group (Early or Middle Jurassic), Tamaulipas, Mexico: Geological Society of America Abstracts with Programs, v. 20, p. 317–318.

Ferguson, M.W.J., 1985, Reproductive biology and embryology of the Crocodilians, in Gans, C., ed., Biology of the Reptilia: New York, Wiley and Sons Inc., p. 329–491.

Ferguson, M.W.J.,a nd Joanen, T., 1982, Temperature of egg incubation determines sex in *Alligator mississippiensis:* Nature, v. 296, p. 850–853.

—— , 1983, Temperature-dependent sex determination in *Alligator mississippiensis:* Journal of zoology, v. 200, p. 143–177.

Findley, R., 1981, St. Helens; Mountain with a death wish: National Geographic, v. 159, p. 2–65.

Francis, P., 1983, Giant volcanic calderas: Scientific American, p. 60–70.

Fridriksson, S., 1975, Surtsey: New York, Wiley and Sons, 198 p.

Fritz, W. J., 1986, Plant taphonomy in areas of explosive volcanism, in Broadhead, T. W., ed., Land plants; Notes for a short course: Knoxville, University of Tennessee Department of Geological Sciences Studies in Geology no. 15, p. 1–9.

Fritz, W. J., and Harrison, S., 1985, Transported trees from the 1982 Mount St. Helens sediment flows; Their use as paleocurrent indicators: Sedimentary Geology, v. 42, p. 49–64.

Froggatt, P. C., Wilson, C.J.N., and Walker, G.P.I., 1981, Orientation of logs in the Taupo ignimbrite as an indicator of flow direction and vent position: Geology, v. 9, p. 109–111.

Gledhill, J. A., 1986, Dinosaur extinction and volcanic activity: EOS (Transactions of the American Geophysical Union), v. 66, p. 153.

Gore, R., 1984, The dead do tell tales at Vesuvius: National Geographic, v. 165, p. 557–613.

Hammer, C. U., 1980, Acidity of polar ice cores in relation to absolute dating, past volcanism, and radio-echoes: Journal of Glaciology, v. 25, p. 359–372.

Hay, R. L., and Leakey, M., 1982, The fossil footprints of Laetoli: Scientific American, v. 246, p. 50–57.

Haymon, R. M., Koski, R. A., and Sinclair, C., 1984, Fossils of hydrothermal vent worms from Cretaceous sulfide ores of the Samail Ophiolite, Oman: Science, v. 223, p. 1407–1409.

Hayward, B. W., 1976, Macropalentology and paleoecology of the Waitakere Group, (lower Miocene) Waitakere Hills, Auckland: Journal of the Auckland University Field Club, v. 22, p. 177–206.

—— , 1977a, Lower Miocene Polychaetes from the Waitakere Ranges, North Auckland, New Zealand: Journal Royal Society of New Zealand, v. 7, p. 5–16.

—— , 1977b, Lower Miocene corals from the Waitakere Ranges, North Auckland, New Zealand: Journal of the Royal Society of New Zealand, v. 7, p. 99–111.

Hayward, J. L., Miller, D. E., and Hill, C. R., 1982, Volcanic ash fallout; Its impact on breeding gulls, in Keller, S.A.C., ed., Mount St. Helens; One year later: Cheney, East Washington University Press, p. 141–142.

Hayward, J. L., Amlaner, C. J. and Young, K. A., 1989, Turning eggs to fossils; A natural experiment in taphonomy: Journal of Vertebrate Paleontology, v. 9, p. 196–200.

Hirsch, K. F., and Lopez-Jurado, L. F., 1987, Pliocene Chelonian fossil eggs from Gran Canaria, Canary Islands: Journal of Vertebrate Paleontology, v. 7, p. 96–99.

Holmes, A., 1965, Principles of physical geology (2nd ed.:) London, Nelson and Sons, 1288 p.

Hooker, J. S., 1987, Late Cretaceous ashfall and the demise of a Hadrosaurian "herd": Geological Society of America Abstracts with Programs, v. 19, p. 284.

Horne, G. S., 1976, Geology of Lower Ordovician fossiliferous strata between Virgin Arm and Squid Cove, New World Island, Newfoundland: Geological Survey of Canada Bulletin 261, p. 1–9.

Horner, J., 1988, Digging dinosaurs: New York, Workman Publishing Co., 210 p.

Hsü, K. J., and MacKenzie, J. A., 1985, A Strangelove ocean in the earliest Tertiary, in Sundquist, T., and Broecker, W., eds., The carbon cycle and atmospheric CO_2: American Geophysical Union Geophysical Monograph Series 32, p. 487–492.

Hunt, A. P., 1987, Late Cretaceous dinosaur distribution patterns; Do we need volcanic island arcs and aseismic ridges to act as "landbridges": Geological Society of America Abstracts with Programs, v. 19, p. 284.

Jain, S. L., 1989, Recent dinosaur discoveries in India, including eggshells, nest, and coprolites, in Gillette, D. D., and Lockley, M. G., eds., Dinosaur tracks and traces: Cambridge University Press, p. 99–108.

Jashemski, W. F., 1979, The gardens of Pompeii, Herculaneum, and the villas destroyed by Vesuvius: New York, Caratzas Brothers, 372 p.

Jefferson, T. H., 1982, The preservation of fossil leaves in Cretaceous volcaniclastic rocks from Alexander Island, Antarctica: Geological Magazine, v. 119, p. 291–300.

Keith, M. L., 1982, Violent volcanism, stagnant oceans, and some inferences regarding petroleum, stratabound ores, and mass extinctions: Geochemica et Cosmochemica Acta, v. 46, p. 2621–2631.

Lee, D. E., 1980, The Cenozoic and Recent Rhychonellide brachiopods of New Zealand with an account of the Eocene and Paleocene brachiopod faunas [Ph.D. thesis]: University of Otago, xxx p.

—— , 1986, Paleoecology and biogeography of the New Zealand Paleogene brachiopod fauna, in Racheboef, P. R., and Emig, C. C., eds., Les Brachiopods fossiles et actuels: Biostratigraphie du Paleozoique, no. 4, 500 p.

Leeder, M., 1982, Sedimentology; Process and product: London, George Allen and Unwin, 334 p.

Licht, E. L., 1986, Araneid taphonomy; A paleothermometer: 4th North American Paleontology Congress, Boulder, Colorado, Abstracts, p. A28.

—— , 1987, Araneid fossils in volcanogenic and non-volcanogenic deposits; A comparison of information lost and found: Geological Society of America Abstracts with Programs, v. 19, p. 314.

Lockley, M. G., 1978, The application of ecological theory to paleoecological studies, with special reference to equilibrium theory and the Ordovician system: Lethaia, v. 11, p. 281–291.

—— , 1984, Faunas in volcanoclastic debris flow from the Welsh Basin; A synthesis of paleoecological and volcanological observations, in Bruton, D. L., ed., Aspects of the Ordovician system: Palaeontological Contributions from the University of Oslo no. 295, Universitetsforlaget, p. 195–201.

Lockley, M. G., and Chesson, R. H., 1987, Volcanism and fossil biotas; How volcanic activity affects the biostratigraphic record: Geological Society of America Abstracts with Programs, v. 19, p. 315.

Lockley, R. M., 1976, A biography of Gilbert White; Author of the natural history of Selborne, 2nd ed.: London, White Lion Publishing, 136 p.

MacGinitie, H. D., 1953, Fossil plants of the Florissant beds, Colorado: Carnegie Institute of Washington Publication 599, p. 1–188.

McCartney, K. and Loper, D. E., 1989, Emergence of a rival paradigm to account for the Cretaceous/Tertiary Event: Journal of Geological Education, v. 37, p. 36–48.

McKee, E., 1979, A study of global sand seas: U.S. Geological Survey Professional Paper 1052, 429 p.

McLean, D. M., 1978, A terminal Mesozoic "greenhouse"; Lessons from the past: Science, v. 201, p. 401–406.

—— , 1981a, A test of terminal Mesozoic catastrophe: Earth and Planetary Science Letters, v. 53, p. 103–108.

—— , 1981b, Size factor in the late Pliocene mammalian extinctions: American Journal of Science, v. 281, p. 1144–1152.

—— , 1985a, Mantle degassing induced dead ocean in the Cretaceous–Tertiary transition, in Sundquist, E. T., and Broecker, W., eds., The carbon cycle and atmospheric CO_2; Natural variations Archean to Present: American Geophysical Union Geophysical Monograph Serial 32, p. 493–503.

—— , 1985b, Mantle degassing unification of the trans-K–T geobiological record: Evolutionary Biology, v. 19, p. 287–313.

—— , 1985c, Deccan traps mantle degassing in the terminal Cretaceous marine extinctions: Cretaceous Research, v. 6, p. 235–259.

—— , 1986, Embryogenesis dysfunction in the Pleistocene/Holocene transitions; Mammalian extinctions, dwarfing, and skeletal abnormality, in McDonald, J. N., ed., The Quaternary of Virginia: Virginia Division of Mineral Resources Publication 75, p. 105–120.

—— , 1987, Thermodynamical control in bioevolution and extinction: Geological Society of America Abstracts with Programs, v. 19, p. 321.

McLeroy, C. A., and Anderson, R. Y., 1966, Laminations of the Oligocene Florissant Lake bed deposits, Colorado: Geological Society of America Bulletin, v. 77, p. 605–618.

Morgan, W. J., 1986, Flood basalts and mass extinctions: EOS Transactions of the American Geophysical Union, v. 67, p. 391.

Naranjo, J. L., Sigurdsson, H., Carey, S. N., and Fritz, W., 1986, Eruption of the Nevado del Ruiz Volcano, Columbia, on 13 November 1985; Tephra fall and lahars: Science, v. 233, p. 961–963.

Neuman, R. B., 1972, Brachiopods of Early Ordovician volcanic islands: Proceedings of the 24th International Geological Congress, Montreal, Canada, v. 7, p. 297–302.

—— , 1976, Early Ordovician (late Arenig) brachiopods from Virgin Arm, New World Island, Newfoundland: Geological Survey of Canada Bulletin 261,

p. 11–61.

——, 1984, Geology and paleobiology of islands in the Iapetus Ocean; Review and implications: Geological Society of America Bulletin, v. 95, p. 1188–1201.

Nopsca, F., 1923, On the geological importance of the primitive reptilian fauna in the uppermost Cretaceous of Hungary: Quarterly Journal of the Geological Society of London, v. 79, p. 100–116.

——, 1934, The influence of geological and other climatological factors on the distribution of nonmarine fossil reptiles and Stegocephalia: Quarterly Journal of the Geological Society of London, v. 60, p. 76–140.

Oelofsen, B. W., 1978, Atmospheric carbon dioxide/oxygen imbalance in the Late Cretaceous hatching of eggs and the extinction of biota: Palaeontologica Africana, v. 21, p. 45–51.

Orth, C. J., Gilmore, J. S., Knight, J. D., Pilmore, C. L., Tschudy, R. H., and Fassett, J. E., 1981, An iridium abundance anomaly in the palynological Cretaceous–Tertiary boundary in northern New Mexico: Science, v. 214, p. 1341–1343.

Oudin, E., and Constantino, G., 1984, Black smoker chimney fragments in Cyprus sulphide deposits: Nature, v. 308, p. 349–352.

Paladino, F. V., Dodson, P., Hammond, J. K. and Spotila, J. R., 1989, Temperature-dependent sex determination in dinosaurs? Implications for population dynamics and extinction, in Farlow, J. O., ed., Paleobiology of the Dinosaurs: Boulder, Colorado, Geological Society of America Special Paper 238, p. 63–70.

Rex, G. M., and Scott, A. C., 1987, The sedimentology and preservation of the lower Carboniferous plant deposits at Pettycur Fife, Scotland: Geological Magazine, v. 124, p. 43–66.

Rice, A., 1987a, Explosive volcanism; A source of shocked minerals at the K/T boundary? [abs.]: EOS Transactions of the American Geophysical Union, v. 67, p. 390.

——, 1987b, Did meteorites really pick off the dinosaurs; Shocking evidence from volcanos: Geological Society of America Abstracts with Programs, v. 19, p. 328.

Sahni, A., 1988, Cretaceous–Tertiary boundary events; Mass extinctions, iridium enrichment, and Deccan volcanism: Current Science, v. 57, p. 513–519.

——, 1989, Paleoecology and paleoenvironments of Late Cretaceous dinosaur eggshell sites from Penninsular India, in Gillette, D. D., and Lockley, M. G., eds., Dinosaur tracks and traces: Cambridge University Press, p. 179–185.

Scott, A. C., 1987, The ecology of Scottish lower Carboniferous floras [abs.]: 11th International Congress of Carboniferous Stratigraphy and Geology, Beijing, China, p. 148–149.

Scott, A. C., and Galtier, J., 1985, Distribution and ecology of early ferns: Proceedings of the Royal Society of Edinburgh, v. 86-B, p. 141–149.

Sear, C. B., Kelly, P. M., Jones, P. D., and Goodess, C. M., 1987, Global surface temperature responses to major volcanic eruptions: Nature, v. 330, p. 365–367.

Sepkoski, J. J., Jr., 1981, A factor analytic description of the Phanerozoic marine fossil record: Paleobiology, v. 7, p. 36–53.

Seymour, R. S., 1979, Dinosaur eggs; The relationship between gas conductance through the shell, water loss during incubation, and clutch size: Paleobiology, v. 5, p. 1–11.

Seymour, R. S., and Ackerman, R. A., 1980, Adaptations to underground nesting in birds and reptiles: American Zoologist, v. 20, p. 437–447.

Sheets, P. D., 1981, Volcanism and the Maya: Natural History, v. 90, p. 32–41.

Sigurdsson, H., 1982, Volcanic pollution and climate; The 1783 Laki eruption: EOS Transactions of the American Geophysical Union, v. 63, p. 601–602.

——, 1987, Lethal gas bursts from Cameroon crater lakes: EOS Transactions of the American Geophysical Union, v. 68, p. 570–573.

Sigurdsson, H., and Carey, S., 1988, The far reach of Tambora: Natural History, v. 97, p. 66–72.

Sigurdsson, H., Carey, S., Cornell, W., and Pescatore, T., 1985, The eruption of Vesuvius in A.D. 79: National Geographic Research, v. 1, p. 332–387.

Sigurdsson, H., Devine, J. D., and Davis, A. N., 1985b, The petrologic estimation of volcanic degassing: Jokull 35, v. 22, p. 1–8.

Sloan, R. E., Rigby, J. K., Van Valen, L. M., and Gabriel, D., 1986, Gradual dinosaur extinction and simultaneous ungulate radiation in the Hell Creek Formation: Science, v. 232, p. 629–633.

Smith, G. A., 1986, Coarse-grained nonmarine volcaniclastic sediment; Terminology and depositional process: Geological Society of America Bulletin, v. 97, p. 1–10.

Sparks, R.S.J., and Wilson, L., 1976, A model for the formation of ignimbrite by gravitational column collapse: Journal of the Geological Society of London, v. 132, p. 441–451.

Spicer, R. A., Burnham, R. J., Grant, P., and Glicken, H., 1985, *Pityogramma calomelanos,* the primary post-eruption colonizers of Volcan Chichonal, Chiapes, Mexico: American Fern Journal, v. 75, p. 1–5.

Stokes, W. L., 1945, A new quarry for Jurassic dinosaurs: Science, v. 101, p. 115–117.

Stommel, H. M., and Stommel, E., 1983, Volcano weather; The story of 1816, a year without summer: New Port, Rhode Island, Seven Seas Press, 177 p.

Suthren, R. J., and Furnes, H., 1980, Origin of some bedded welded tuffs: Bulletin of Volcanology, v. 43, p. 61–71.

Taggert, R. E., and Cross, A. T., 1987, Plant successions and interruptions in Miocene volcanic deposits: Geological Society of America Abstracts with Programs, v. 19, p. 338.

Thorarinsson, S., 1969, The Lakagigar eruption of 1783: Bulletin of Volcanology, v. 33, p. 910–927.

Tilling, R. I., Rubin, M., Sigurdson, H., Carey, S., Duffield, W. A., and Rose, W. I., 1984, Holocene eruptive activity of El Chichon volcano, Chiapas, Mexico: Science, v. 224, p. 747–749.

Tuttle, M. L., and 10 others, 1987, The 21st August 1986 Lake Nyos gas disaster: U.S. Geological Survey Open-File Report 87–97, 58 p.

Vitousek, P. M., Walker, L. R., Witeaker, L. D., Mueller-Dombois, D., and Matson, P. A., 1987, Biological invasion by *Myrica faya* alters ecosystem development in Hawaii: Science, v. 238, p. 802–804.

Voorhies, M. R., 1981, Ancient ashfall creates a Pompeii of prehistoric animals: National Geographic, v. 159, p. 66–75.

——, 1987, Terrestrial vertebrate assemblages preserved in volcanic ash and sediments; Comparisons from the late Miocene of Nebraska: Geological Society of America Abstracts with Programs, v. 19, p. 338.

Voorhies, M. R., and Thomasson, J. R., 1979, Fossil grass *Anthoecia* within Miocene rhinoceros skeletons; Diet in an extinct species: Science, v. 206, p. 331–333.

Ward, P., Wiedmann, J., and Mount, J. F., 1986, Maastrichtian molluscan biostratigraphy and extinction patterns in a Cretaceous/Tertiary boundary section exposed at Zumaya, Spain: Geology, v. 14, p. 899–903.

Weintraub, B., 1982, The disaster of El Chicon: National Geographic, v. 162, p. 654–684.

Weishampel, D. B., and Reif, W. E., 1984, The work of Franz Baron Nopcsa (1877–1933); Dinosaurs, evolution, and theoretical tectonics: Jahrbuch der Geologischen Bunde sanstalt wien, v. 127, p. 187–203.

Whaley, P., 1987, Insects and Cretaceous mass extinction: Nature, v. 327, p. 562.

White, G., 1936, The natural history of Selbourne: New York, Methuen and Co., 227 p.

Williams, A., and others, 1972, A correlation of Ordovician rocks in the British Isles: Geological Society of London Special Report 3, p. 1–74.

Williams, D.L.G., Seymour, R. S., and Kerouio, P., 1984, Structure of fossil eggshell from the Aix Basin, France: Paleogeography, Paleoclimatology, Paleoecology, v. 45, p. 23–37.

Wood, C. A., 1984, The amazing and portentous summer of 1783: EOS Transactions of the American Geophysical Union, v. 65, p. 410.

Zoller, W. H., Parrington, J. R., and Phelon Kotra, J. M., 1983, Iridium enrichment in airborne particles from Kilauea Volcano; January 1983: Science, v. 222, p. 1118.

MANUSCRIPT ACCEPTED BY THE SOCIETY JUNE 21, 1989

Printed in U.S.A.

An early terrestrial biota preserved by Visean vulcanicity in Scotland

W.D.I. Rolfe
Keeper of Geology, National Museums of Scotland, Chambers Street, Edinburgh EH1 1JF, Scotland
G. P. Durant
Hunterian Museum, University of Glasgow, Glasgow G12 8QQ, Scotland
A. E. Fallick
Scottish Universities Research and Reactor Centre, East Kilbride, Glasgow G75 0QU, Scotland
A. J. Hall
Department of Geology and Applied Geology, University of Glasgow, Glasgow G12 8QQ, Scotland
D. J. Large
Department of Earth Sciences, University of Cambridge, Downing Street, Cambridge CB2 3EQ, England
A. C. Scott
RHB New College, University of London, Egham, Surrey TW20 0EX, England
T. R. Smithson
Biology Department, Cambridge Regional College, Newmarket Road, Cambridge CB5 8EG, England
G. M. Walkden
Department of Geology and Mineralogy, Marischal College, University of Aberdeen, Aberdeen AB9 1AS, Scotland

ABSTRACT

An unusual, laminated, spherulitic limestone with cherty layers forms part of a volcanogenic sequence in the Midland Valley of Scotland, 27 km west of Edinburgh. It preserves a cross section of the early Carboniferous terrestrial community. Microcrystalline silica laminae containing inclusions of calcite or dolomite may be primary, and support a hot-spring origin for the deposit. Oxygen isotope analyses of the silica and carbonate are consistent with the precipitation of silica from hot, possibly boiling, hydrothermal solution. Such hot spring waters were presumably heated by hypabyssal intrusives associated with the West Lothian volcanic center, only 5 km to the northwest. Silica was probably precipitated by acidification when such waters entered a very local, fresh-water lake. At the same time, calcium carbonate was precipitated from surface water, due to an increase in temperature and reduction in acidity. Such precipitates covered wide areas of the lake floor and the fossils lying on it. Algal or bacteriogenic precipitation may have been responsible for the formation of accretionary growths around nuclei of wood, other fossils, and rock clasts.

The preserved terrestrial biota includes almost-complete individuals of a reptile and of four amphibian groups: temnospondyls, anthracosaurs, loxommatids, and aïstopods. These are the oldest known, fully land-going tetrapods. Truly aquatic forms are absent. Invertebrates include: large eurypterids, which may have been partially terrestrial; the oldest known proven terrestrial scorpion (*Gigantoscorpio*); the earliest harvestman "spider"; and millipedes. Several land-plant assemblages are found in the sequence. Permineralized plants show exceptional anatomical preservation when enclosed within

accretionary nodules. Fusainized (charred) gymnosperm wood and pteridosperm leaves indicate the presence of wildfires.

Fully articulated amphibian skeletons occur within the laminated limestones, whereas only disarticulated skeletons have been collected from the lapilli tuffs. Such taphonomic evidence supports an epiclastic origin for the volcaniclastic rocks, deduced from study of clasts and bedform analysis. Fragments mass-flowed or were rain-washed from the flanks of a small basaltic volcano onto or into an area of sinter deposition.

INTRODUCTION

The East Kirkton site in the Bathgate Hills, 27 km west of Edinburgh, has long been recognized as a volcanogenic sequence showing hot-spring features (Hibbert, 1836; Geikie, 1861; Cadell, 1925; Muir and Walton, 1957). Only a few fossils were previously known from this site. The most spectacular of them were "Scouler's heids"—rare heads of the unusual eurypterid *Hibbertopterus scouleri* (Hibbert, 1836). Abundant plants were also known from shales near the top of the sequence. In 1984, Wood discovered a wealth of previously unknown terrestrial forms in dry-stone wall slabs taken from this quarry some time before its closure in 1844 (Wood and others, 1985; Cadell, 1925). These included very rare, nearly complete specimens of four major groups of fossil amphibians, including the earliest known complete amphibian ancestors of frogs and salamanders, and of reptiles (Milner and others, 1986; Rolfe, 1986b; Wood and others, 1985). Since most previously known early fossil amphibians were aquatic forms that lived in coal swamps, these new finds alter the current scenario of terrestrial amphibian evolution. The site provides evidence of several fully land-going amphibians at an age much earlier than had previous been recognized. In 1988, Wood found the oldest-known reptile at this site (Gee, 1988). The peculiar volcanic conditions prevailing at that time have preserved several other land animals besides the tetrapods. These include the earliest known harvestman "spider," millipedes, scorpions, and fragments of large eurypterids. Before this overwhelmingly terrestrial assemblage had been found, it had been argued from quite different, morphological grounds (Rolfe, 1986a, p. 311, but paper delivered in 1979) that the eurypterid *Hibbertopterus* showed terrestrial adaptations. This, however, remains controversial (Milner and others, 1986, p. 25; Selden, 1985; Waterston and others, 1985). Conversely, such early scorpions have been recently considered to be largely aquatic (Kjellesvig-Waering, 1986). Current study by A. Jeram indicates that a large specimen of one of these *Gigantoscorpio* scorpions probably bears terrestrial respiratory organs, thereby making it the oldest known terrestrial scorpion (Selden and Jeram 1989, p. 305).

The East Kirkton assemblage has international significance, therefore, as a rare, early terrestrial biota preserved by volcanogenic processes. A team of 38 research workers has recently been assembled to work on the material collected to date. Therefore, only a preliminary report is given here, and many problems remain to be solved. A bed-by-bed excavation was started in 1985. In 1987, the National Museums of Scotland took up this project, sampled the top half of the sequence in detail, and sank two exploratory bore holes. The remainder of the sequence and its lateral variation will be investigated by future excavations. Authorship of sections in this chapter is as follows: Sequence, Rolfe and Durant; Hydrothermal silica genesis, Hall, Large, and Fallick; Carbonate laminites and diagenesis, Walkden; Plants, Scott; Vertebrates, Smithson; Volcanic setting, Durant.

SEQUENCE

The so-called East Kirkton Limestone is an upper Visean complex of spherulitic limestones, cherty limestones, and cherts separated by bands of tuff occurring within a thick sequence of olivine basaltic lavas and tuffs, near the top of the Upper Oil Shale Group. It is laterally impersistent and may be confined to little more than the area of the quarry in which it is exposed, i.e., about 400 to 600 m in lateral extent. The sequence has yet to be described in formal stratigraphic detail, but a summary follows of the generalized section shown in Figure 1.

Basal massive limestones and tuffs

The base of the sequence, as currently exposed, comprises a massive limestone (probably bed c of Geikie, 1861). This bed varies in thickness from 1.4 m to 2.3 m within a distance of 5.5 m (Muir and Walton, 1957, p. 158, Fig. 2a). The limestone displays accretionary textures with abundant vugs filled with chalcedony, quartz, carbonate, and bitumen. Large (1987) believes that this limestone may represent a carbonated siliceous sinter (see next section).

Tuffs and limestones

Approximately 0.5 m of calcareous tuff lying on top of the basal limestone contains limy pebbles and nodular lenticles of limestone. Thin limestones, tuffs, and shales with chert compose the next 400 mm of the section. Occasional isolated amphibian bones are found in the coarser tuffs. The whole series of units probably corresponds with beds d through i of Geikie (1861).

Laminated limestone

A 4.2 to 5-m-thick spherulitic limestone follows, characterized by fine laminae less than 1 mm thick (bed k of Geikie, 1861). (Detailed consideration of the origin of this unit is given in the

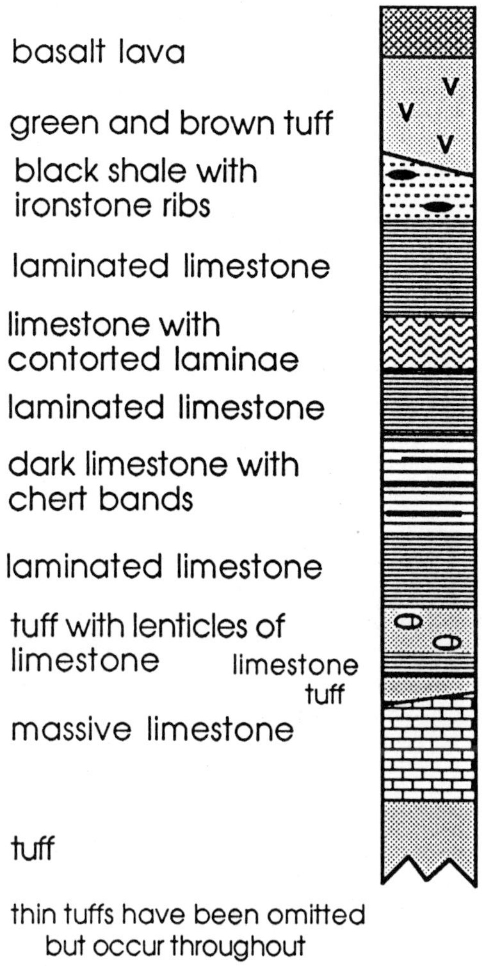
basalt lava
green and brown tuff
black shale with ironstone ribs
laminated limestone
limestone with contorted laminae
laminated limestone
dark limestone with chert bands
laminated limestone
tuff with lenticles of limestone
limestone
tuff
massive limestone
tuff
thin tuffs have been omitted but occur throughout

Figure 1. Generalized section at East Kirkton (not to scale).

following section.) The laminae are sometimes strongly deformed (Fig. 2). Unpublished analysis of the vergences of these structures by J. Cater suggests that the upper 2 m of this unit and, perhaps, the overlying volcaniclastics and lava, moved SE–ESE as a semi-coherent block over the lower part of the unit. This movement, Cater suggests, occurred prior to deep burial and lithification, but after some compaction and dewatering. The slide may have been triggered by volcanic/seismic instability due to inflation of the West Lothian volcanic center 5 km to the northwest, prior to eruption.

Part of the thickness of the laminated limestone may be locally occupied by up to 1.5 m of massive nodular limestone. It is the laminated limestone proper that contains the complete amphibians, lying on the bedding-plane surfaces, together with well-preserved plants, eurypterids, scorpions, and a harvestman spider.

Black shale with ironstone

Black shales (bed m of Geikie, 1861) overlie the laminated limestone, with thin, clayband ironstone bands and nodules, containing fragmental calcareous horizons. These shales are of very local occurrence and are lenticular, reaching a maximum thickness of 1.6 m. This horizon has long been known to yield plants (see below), but is now known also to yield rhizodont fish (and probably also acanthodians and a shark, *Tristychius*), as well as many ostracods, scorpion and eurypterid cuticle, and occasional bivalves. Preliminary studies suggest this is a nonmarine shale.

Tuffs

Fine and coarse green basaltic tuffs (bed n of Geikie, 1861) as thick as 4.2 m contain abundant roots toward the top of the sequence. The water-lain nature of this deposit is confirmed by the presence of usually isolated palaeoniscoid fish scales and occasional ostracods. This detritus probably washed off the flanks of the nearby volcano. The top of the tuff shows coarse angular rock fragments and rare rounded boulders of lava (bombs?). The sequence is capped by a fine-grained olivine basalt lava, locally seen to occupy channels in the tuff.

HYDROTHERMAL SILICA

One model being developed from examination of material derived from the lowest part of the sequence is summarized in this section.

The clearest evidence of chemical precipitation is the finely laminated nature of chert and carbonate. Silica and carbonate, however, display many intricate spherulitic, mammillary, and banded textures of variable grain size (see following section). A careful search through siliceous material did not reveal any base-metal sulfides or exotic minerals such as metals, alloys, sulfides, and sulfosalts of Au, Ag, Cu, Hg, Sb, Bi, and As, which might be expected to occur in a synvolcanic epithermal-hydrothermal environment (Berger and Bethke, 1985). It is noteworthy that native silver associated with niccolite is recorded about 2 km to the north at the old Hilderston silver mine (Stephenson and others, 1983). Pyrite is found in the siliceous carbonate lithologies, but it is rare and occurs as scattered aggregates of grains and framboids often associated with fossil cellular material. Pyrite is more abundant as fine-grained aggregates and disseminations in nodules or layers in shales within the sedimentary sequence, where it is presumably of bacteriogenic origin.

The highest concentrations of silica occur in massive, coarsely crystalline, pale brown limestone (Fig. 1, massive limestone), which is permeated with several forms of silica. Dark and white chalcedony, quartz, calcite, and bitumen are commonly found filling vugs in this limestone. A crystallization sequence is recognized in the vugs, with initial chalcedony followed by coarse clear quartz followed by coarse clear calcite. Bitumen is an occasional final-stage infill and occurs as globules and stains on earlier phases. Coarse calcite and bitumen are found more frequently in late cross-cutting veins, often tension fractures, in brittle rocks in the sequence. They are interpreted as being related to late post-lithification liquid movement.

The carbonate, which hosts the chalcedony-lined vugs, generally has a spheroidal (cm-scale) carbonate texture with convex-in margins to vugs, indicating that the vugs represent space interstitial to the interfering spheroidal mass of the limestone. Relict laminated cherty layers in this lithology and clear textural evidence of laminated chert being replaced by carbonate suggests some of the textural complexity of this lithology results from dissolution, remolibization, and reprecipitation of carbonate and silica by hydrothermal solutions percolating through a carbonate/silica mound. Preliminary oxygen isotope analyses of silica and carbonate, assuming that the water was either early Carboniferous marine or meteoric (both would have O^{18} = 0 ppm, approximately), give the following results:

matrix calcite	23.5 ppm	~50°C
white chalcedony	25.0 ppm	~70°C
dark chalcedony	23.1 ppm	~80°C
clear quartz	21.5 ppm	~95°C
white laminated chert	23.2 ppm	~80°C

Uncertainty in the water $\delta^{18}O$ value of ± 4 ppm leads to temperature uncertainties of about ± 25°C. If the water had re-equilibrated with igneous rock at depth and inherited a magmatic oxygen isotope signature of +8 ppm, then the calculated temperatures would range from 130° to 170°C. Unfortunately, no fluid inclusions could be found in several polished wafers of coarse quartz. The calcite of the host limestone is clearly not in equilibrium with any of the varieties of silica; the temperature sequence parallels the paragenetic sequence, with the coarse quartz crystallizing late from the highest-temperature fluid. The isotopic analyses are therefore consistent with precipitation of silica from hot, possibly boiling, hydrothermal solution.

The East Kirkton sequence lies only about 5 km southeast of the lower Carboniferous West Lothian volcanic center (Cadell, 1925). The volcanics are dominantly alkaline in composition. The heat for the hydrothermal system could well have been provided by magma in feeders or sills at depth; alkali dolerite (teschenite) intrusives are generally associated with early Carboniferous volcanic activity in the Midland Valley of Scotland (Francis, 1983). Interaction of water, especially meteoric, and alkaline basaltic rocks at moderate temperature would have promoted zeolitization of basalt, increase in alkalinity of the liquid, and mobilization of silica (Fournier, 1985). On entering the freshwater lake environment of the East Kirkton sequence, hot siliceous alkaline water would precipitate silica by acidification. The near-surface calcium bicarbonate-bearing lake water would then,

Figure 2. Penecontemporaneous convolution of the laminated limestone, showing a SE-verging fold, indicating movement of upper semi-coherent strata over lower. (Drawing by Mark Fothergill.)

in turn, precipitate calcium carbonate due to an increase in temperature and a decrease in acidity. Therefore, the scene is set for chemical precipitation of silica and carbonate over a wide area of the lake floor as well as on and within siliceous carbonate mounds. It is also likely that organic (algal or bacteriogenic) processes that could also lead to carbonate precipitation would have been enhanced in this environment. Botryoidal masses with laminated internal layering and stromatolitic-like texture could, therefore, represent either inorganic or biochemical accretion of carbonate, or both. It is clear from thin-section petrography that millimeter-sized pellets of accreted carbonate were also formed, transported, and then deposited in the laminated rocks (Large, 1987). Hydrothermal solutions may have passed through permeable horizons en route to the surface and led to "early diagenetic" silicification and preservation of organic material.

Such a syn-depositional hydrothermal input of silica into the East Kirkton volcano-sedimentary sequence, and the concomitant enhancement of carbonate precipitation, explains the unusual variety of textures observed in the siliceous limestone, and the early silicification and carbonation of fossil material.

CARBONATE LAMINITES

Description of the sediments

A particular characteristic of the upper limestones (Fig. 1) is the presence of millimeter-scale laminations. These are defined by dark clay partings and consist of a variety of lithologies, including calcite silt, calcite allochems, calcite spheroids, dolomite, chert, and volcanic ash. A few laminae are graded, and these commonly contain spheroids, sometimes broken. There is great variation in thickness of laminae, and some are actually thin beds 10 mm or more thick. Individual laminae and groups of laminae are laterally persistent and may permit local correlation.

Calcite spheroids

The most characteristic component of the laminites is the spheroid. Typically these are spherical or subspherical aggregates that are 0.1 to 1 mm in diameter. They are made up of radial calcite fibers, occasionally with growth banding. Spheroids occur individually along laminae or as accumulations set in a matrix of carbonate silt, dolomite, or mud. Two types have been recognized. Type 1 spheroids consist of generally inclusion-free, brown, fibrous calcite and ferroan calcite with a sharply defined pseudouniaxial cross under cross polars. These show sharp and sometimes very fine growth banding (microzonation) in cathodoluminescence. Type 2 spheroids consist of inclusion-rich, brown, fibrous calcite with the less well-defined pseudouniaxial cross. These spheroids show broad patchy growth zones in cathodoluminescence. Inclusions in the spheroids include microdolomites and indeterminate allochems, but more rarely there are cylindrical objects that are apparently biogenic.

These cylindrical objects, about 5 μm in diameter, could be interpreted as broken segments of calcified fibers of algal or bacterial origin. Tubes of similar diameter, but more complete, are commonly associated with type 1 spheroids. Less well-defined tubes occasionally extend perpendicular to bedding across a lamina, as if in growth position. Other biogenic allochems, consisting of cellular material of algal or higher plant origin, occasionally occur at the cores of larger spheroids. Others are sometimes cored by platey, peloidal, or clotted micritic carbonate.

Spheroids generally exhibit a single generation of calcite fibers. Larger composite growths consisting of many generations also occur. These comprise hemispheres of type 1 calcite, of similar crystal size to that seen in spheroids, stacked into botryoidal masses. They commonly extend along laminae, but the larger ones form lenticular bodies tens or hundreds of millimeters high, against which laminae are almost always compacted and commonly fail.

Origin of spheroids and related calcite

Spheroids are structurally comparable with calcite spherules described from a Mesozoic hot-spring deposit by Steinen and others (1987). The Mesozoic spheroids were smaller, however, and were believed to have precipitated rapidly in suspension through flash boiling of supersaturated water. Steinen and others (1987) also noted the existence of possible algal or bacterial structures, including clotted textures similar to those occasionally seen at the cores of East Kirkton spheroids. None of the other tufa and travertine textures they noted have been identified within the laminated beds.

Spheroids and botryoidal cements are also known to be associated with modern and ancient stromatolites (Dill and others, 1986; Monty, 1976). Similar textures have been noted in Carboniferous sediments from normal marine environments elsewhere in Europe (Oberst, 1987), but spheroids from these situations are much smaller than the East Kirkton examples and rarely associated with filaments.

The East Kirkton spheroids and botryoidal masses are obviously benthonic, and whereas the larger masses grew in situ (Fig. 3), the spheroids probably grew under conditions of periodic or continuous movement comparable with either marine or hot-spring ooids. Type 1 spheroids and their larger equivalents show no sign of recrystallization during diagenesis, and their present characteristics probably reflect their original nature. Type 2 spheroids could have undergone recrystallization, however, based on their relatively poor crystal organization and patchy luminescent banding. If they had originally been a high-Mg calcite, it might account for the presence of dolomite inclusions, but it is recognized that dolomite might have been an original component of the sediment from which the other inclusions in type 2 spheroids were derived.

The presence of graded beds and graded laminae containing spheroids, commonly broken, indicates that some spheroids were being precipitated elsewhere, perhaps in the proximity of a hot spring, and then brought into deeper water. However, the pres-

ASPECTS OF DIAGENESIS

Sparry and fibrous ferroan calcite cements are present, and both are uniformly dull in cathodoluminescence. The fibrous cement is found within clay layers and displaces matrix. Chalcedony is common, replacing carbonate laminae, parts of spheroids, or botryoidal masses. This patchily developed chalcedony is always secondary. Some uniform microcrystalline isotropic silica that forms individual laminae could be primary, but there is no evidence to suggest that silica has been extensively replaced by calcite.

SELECTED BIOTA

Some of the invertebrate faunas have been summarized above, by Rolfe (1988) and in Milner and others (1986). Preliminary accounts of other elements of selected parts of the biota follow.

Plants

Plant fossils have been found loose and in situ in the quarry from most levels in the sequence. There are, however, distinct assemblages occurring within the major lithotypes and a diversity of preservation states. Plant fossils may be preserved as compressions, anatomically preserved as cellular permineralizations, or as fusain (charcoal; Schopf, 1975).

Ash assemblage. The green volcanic ashes at the top of the quarry sequence mainly yield compressions, ranging in size from a few millimeters to 30 cm. They are mostly lycopod stem fragments including *Lepidodendron* sp. (Fig. 4A) and *Lepidophloios* sp. Isolated sporophylls and leaves are also common. Toward the top of the ash sequence, as exposed in the quarry, *Stigmaria* rootlets are abundant in situ. In addition to the lycopods, pith casts of sphenopsids occur. Some fusain is also present in layers containing fragmentary plant material. Megaspores of heterosporous lycopods are abundant at some levels. In palynological preparations, spores are light brown and well preserved. In general aspect, the plants at this level at East Kirkton bear striking resemblance, in terms of preservation, to those from the green ashes at the top of the Loch Humphrey Burn volcanogenic sequence in the Western Midland Valley (Scott and others, 1984). As yet, no permineralized plants have been found at this level, and the flora is dominated by lycopods rather than pteridosperms.

Black shale assemblage. Plant compressions, usually very fragmentary but occasionally larger, occur in the black shales and associated nodules below the ashes. At this level the flora is dominated by frond fragments, mainly of *Sphenopteris* including *S. affinis* Lindley and Hutton, widely known elsewhere at this stratigraphic level (Scott, 1986). Sphenopsid pith casts are found rarely, and lycopods including leafy shoots and cones occur at some levels.

Limestone assemblage. Palaeobotanically, and also preservationally, the most interesting plant fossils occur in the lime-

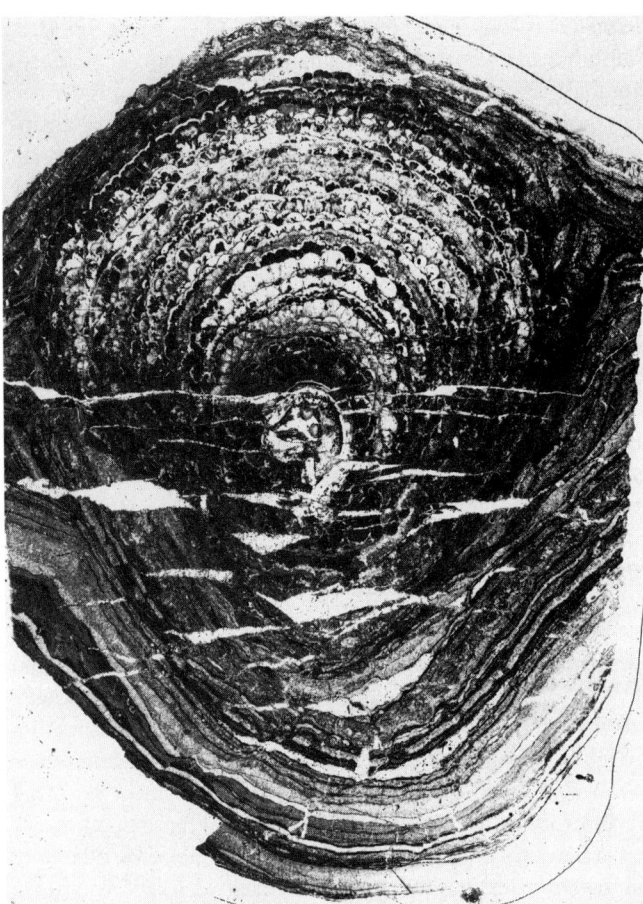

Figure 3. "Stromatolitic" accretionary mass of calcite (diameter about 3 cm) around plant (shown in detail in Fig. 4E) in syngenetically brecciated and deformed laminated limestone.

ence of identical calcite in the autochthonous botryoidal masses clearly suggests that precipitation in situ was also possible. The localized occurrence of these botryoidal masses also demonstrates that precipitation conditions were not uniform across the sediment surface so that the influence of some sort of organic catalyst can be invoked. The presence of filaments within the spheroids may provide evidence for such a catalyst, but equally the filaments may merely have provided convenient substrates for calcite nucleation.

Microzonation, seen especially in type 1 spheroids, clearly indicates fluctuations in precipitation conditions. It is noticeable that zonation appears in clusters, suggesting the fluctuation was regular. The botryoidal growths examined so far show less microzonation, but there is a strong tendency for an increase in brightness toward the edge of each hemisphere. These fluctuations could be attributable to thermal or geochemical variations, but the fine microzonation might be diurnal. Microzonation patterns used as stratigraphic markers reveal that different generations of spheroids have become mixed along individual laminae.

Figure 4. Fossil plants from the late Visean of East Kirkton (scale 1 cm unless otherwise stated): A. *Lepidodendron* sp. Compression in green ashes at top of sequence (ash assemblage) (Glasgow City Museum C 1986-152-1). B. *Sphenopteris* sp. Compression of pteridosperm foliage, partially permineralized (National Museums of Scotland, GY 1985-4-36). C. *Sphenopteris* sp. Compression of pteridosperm foliate (Glasgow City Museum, G 1985-232-45). D. New permineralized gymnosperm stem at center of "stromatolitic" nodule (limestone assemblage) (National Museums of Scotland, GY 1985.4). E. New plant, anatomically preserved by calcareous permineralization at center of "stromatolitic" nodule (limestone assemblage) (20×).

stone in the lower half of the section. Material has been found both in situ in the quarry, in loose piles, and from the walls in the surrounding area. Fusainized fragments occur commonly. These may be of wood, but leaves also have been found belonging to various pteridosperm foliage taxa. There is a wide gradation of preservation states at this level. Large trunk and frond compressions occur, mostly of gymnospermous origin. At some levels, compressions of large fronds of fern-like foliage occur that include pteridosperm and probably true fern foliage. Most abundant are the genera *Sphenopteridium, Sphenopteris,* and *Rhodea.* Whereas the plants often appear only as compressions, many are partially permineralized by calcite or silica (Fig. 4E). In these cases, the leaves may be permineralized (Fig. 4B). Rachises may be anatomically preserved. Although fronds are often compressed, they may retain their original three-dimensional character.

Isolated permineralized stems and rachides also occur at the center of "stromatolitic" nodules (Figs. 3 and 4D). Often these show excellent anatomy (Fig. 4E) and may be permineralized first by calcite and later by silica. In all cases, the original cell walls are preserved. Larger permineralized logs also occur. In these cases, the wood has often been deformed, indicating only a partial permineralization followed by later compression.

Interpretation. Differences in the assemblages may result from various factors, including original ecology, transport, deposition, and preservation. The change in diversity through the sequence may reflect increasing fluctuation of environment due to inundation by thick ashes that allowed only a few species to survive.

Anatomically preserved plants occur widely in volcanogenic sediments of lower Carboniferous age in the Scottish Midland Valley (Scott and Rex, 1987). In many cases, the occurrence of carbonate permineralizing fluids has been attributed to the weathering of basaltic lavas and tuffs. Secondary silicification has also been reported in slightly older rocks at Pettycur in Fife (Rex and Scott, 1987). The occurrence of partially permineralized compressions in limestones and stems at the center of "stromatolitic" nodules is unique in the Midland Valley. The occurrence of fusain indicates wildfires with the production of charcoal (Cope and Chaloner, 1985), which is not surprising given the geologic setting. Some stems appear partially charred. There is no direct evidence of seasonality from growth rings in the wood. Some partial ring-like structures in the cortex of one new gymnosperm have yet to be explained. The occurrence of small rings, perhaps produced locally by water stress, are known from other volcanic settings in Fife (Scott and others, 1986).

The flora of East Kirkton is of particular significance in that some plants combine several preservation states, which will allow correlation of taxa previously known only separately, as compressions or as permineralizations (Galtier, 1986).

Vertebrates

A variety of vertebrate fossils have been found at the quarry. Fish remains have not been recorded from the East Kirkton Limestone, but shark spines, the scales of palaeoniscoids and rhizodonts, and a tiny acanthodian have been found in the shales and ironstones overlying the limestone. In addition, palaeoniscoid scales have been found in the tuff that caps the sequence. Tetrapod remains are thought to occur throughout the entire section of East Kirkton Limestone. Their distribution will be better known following the 1988 excavations. So far, most amphibian specimens have been found in isolated limestone blocks from the local farm walls or the quarry spoil heaps, but in 1985 some specimens were collected in situ near the base of the sequence.

The amphibians are preserved in varying degrees of completeness, ranging from articulated skeletons to isolated bones, and the quality of preservation varies in the different rock types. Isolated elements and complete skeletons occur together in the same rock types. We do not yet know whether they occur together on the same bedding plane, and so cannot be certain if they were preserved under the same conditions. A large number of individual bones have also been found in interbedded tuff at the base of the sequence.

The most common tetrapods found in the limestone are temnospondyls, the Palaeozoic antecedents of living amphibians (Milner, 1988). Two species have been recognized (Wood and others, 1985; Milner and others, 1986). One of these, a *Dendrerpeton*-like form (Milner, 1985), is the most common amphibian in the fauna (Fig. 5). Anthracosaurs, a group of Carboniferous amphibians generally thought to be closely related to amniotes (reptiles, birds, and mammals) (Panchen and Smithson, 1988), are also represented by at least two species, an *Eoherpeton*-like form (Fig. 6) and a proterogyrinid (Milner and others, 1986). The limbless aïstopods are represented by one nearly complete and two incomplete specimens, one collected in situ from the base of the sequence. Another major group of Carboniferous amphibians, the loxommatids, characterized by peculiar keyhole-shaped orbits, may be represented by an incomplete skull and associated anterior post-crania (Milner and others, 1986).

The anthracosaurs, loxommatid, and temnospondyls at East Kirkton are the earliest known record of each group, but the aïstopods have been recorded in the slightly older Wardie shales (Wellstead, 1982). The rarest tetrapod is the oldest known reptile, 40 m.y. older than previously recorded, of which one specimen is known (Gee, 1988; Smithson, 1989).

The components of the vertebrate fauna at East Kirkton and the state of preservation of individual specimens provide important data by which to interpret the depositional environment of the limestone.

Prior to the discovery of amphibian fossils at East Kirkton, the principal tetrapod sites in the Scottish Mississippian were the lacustrine deposits at Cowdenbeath, Gilmerton, and Loanhead (Smithson, 1985; Milner and others, 1986). There the dominant amphibians are adelogyrinids, anthracosaurs, colosteids, *Crassigyrinus,* and *Doragnathus.* Together they form minor components of a vertebrate fauna dominated by fishes. At East Kirkton, fishes have yet to be found in the limestone, and the only group of

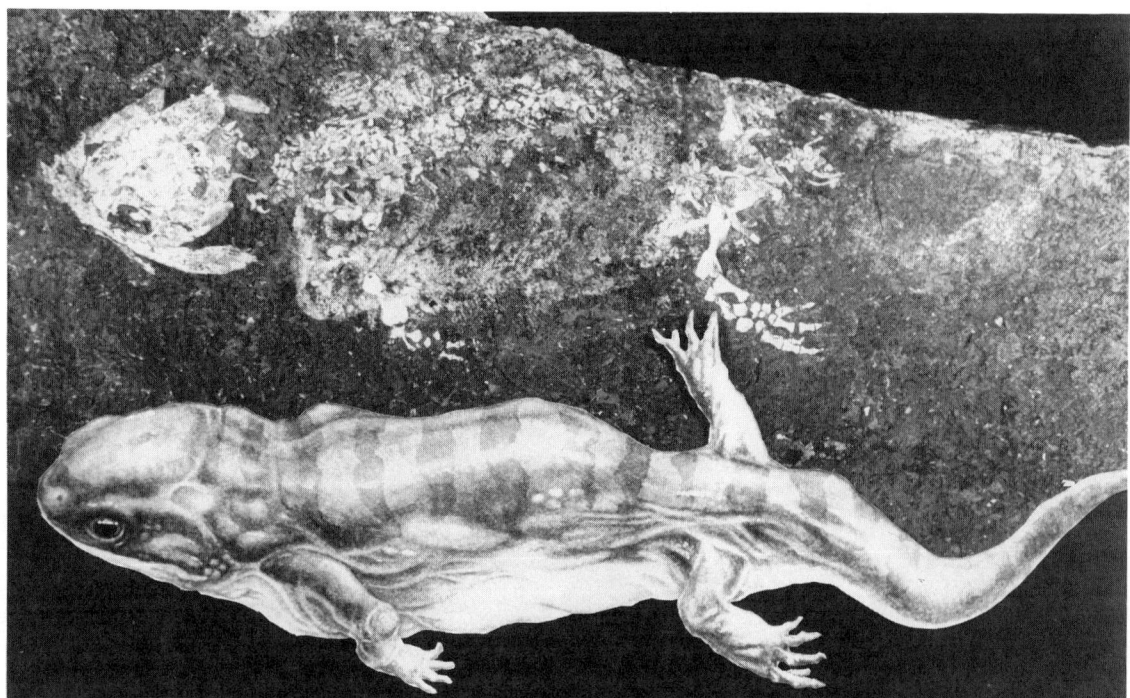

Figure 5. Earliest-known ancestor of frogs and salamanders, a 300-mm-long temnospondyl from the laminated limestone at East Kirkton. The reconstruction shows how the animal probably lay in death. (After Rolfe, 1986b; © Hunterian Museum/*The dinosaur times;* National Museums of Scotland, GY 1985.4.1).

amphibians definitely known to be common to this and the three main lacustrine sites are the anthracosaurs.

The remains of the anthracosaurs and temnospondyls in the East Kirkton Limestone are well ossified. Members of both groups have well-developed appendicular skeletons with long limb bones, ossified tarsals in the ankle, and remarkably, in some of the temnospondyls, ossified carpals in the wrist. In addition, there is no evidence of a lateral line system (an aquatic pressure-sensing system present in fishes and many Palaeozoic tetrapods) in any specimens, and the temnospondyls appear to have a well-developed otic notch, which probably housed a tympanic membrane used for atmospheric hearing. All these features indicate that the anthracosaurs and temnospondyls found at East Kirkton were terrestrial animals.

The absence of fishes and members of the typical Scottish Dinantian tetrapod assemblages (e.g., adelogyrinids, *Crassigyrinus,* and *Doragnathus*) suggest that either the environment of deposition was hostile to their survival or that it was isolated from larger water bodies and the opportunity to colonize was prohibited. The absence of significant numbers of small immature tetrapods in the limestone also suggests that the pools were not amphibian breeding sites. Only one immature temnospondyl has been found. It has a skull approximately 20 mm long, which is half the size of most temnospondyl skulls. This, coupled with the absence of typical freshwater invertebrate fossils (e.g., conchostracans), indicates that the pools were probably unable to sustain a macrofauna. The fossils that have been found, therefore, presumably represent members of a terrestrial community that died in or near the pools (Fig. 7).

Since collecting began in 1984, approximately 20 articulated temnospondyl skeletons have been found, together with skeletons of two anthracosaurs, incomplete remains of three aïstopods, and one loxommatid(?) skull and postcrania. More than 40 isolated bones or restricted associations of elements (e.g., iso-

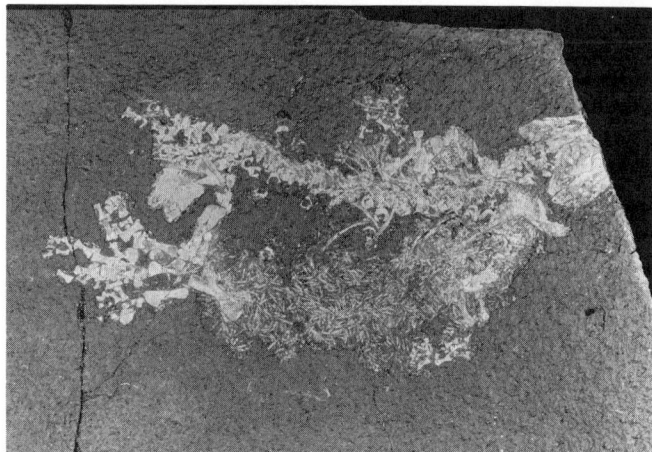

Figure 6. Oldest-known "complete" ancestor of amphibians and reptiles: an almost-complete *Eoherpeton*-like anthracosaur skeleton 214 mm long (National Museums of Scotland, GY 1986.39).

Figure 7. Preliminary reconstruction of the hot springs environment that produced the East Kirkton Limestone and its fossils. The temnospondyl is depicted in the left foreground, the eurpyterid *Hibbertopterus* in the center-right foreground with an *Eoherpeton*-like anthracosaur behind it, next to a fumarole. A harvestman spider is visible at the base of a sphenopsid axis. The pteridosperm *Sphenopteris* is depicted in the bottom-left corner. (After Milner and others, 1986; © Hunterian Museum/*Modern geology*.)

lated skulls, pelvic girdles, and compound vertebrae) have been found in the limestone, and 60 isolated bones in the interbedded tuff.

The occurrence of articulated amphibian skeletons in the limestone suggests rapid burial in a low-energy environment with no benthic organisms disturbing the carcasses. The estimated length of most of the amphibians represented by these skeletons ranges from 400 to 600 mm. Tails are missing from most specimens, however, and the average length of the precaudal skeleton is 250 mm. Larger tetrapods are represented by isolated elements or a few associated bones. Judging by material from elsewhere, these animals had an overall length of 1.5 to 2 m. Remains of the larger members of the fauna are much rarer than those represented by average-sized material. The absence of large articulated skeletons indicates that complete burial did not occur before their carcasses disintegrated and were then dispersed, possibly by water movements generated by streams or hot springs feeding the pools. It is as yet unclear whether the isolated bones found in the interbedded tuff were washed in along with the ash or were originally lying on the pool bottom and were disturbed when the ash "rained in."

VOLCANIC SETTING

Volcanic activity was widespread throughout the Midland Valley of Scotland in Carboniferous times. The area was a major continental alkaline province characterized by a variety of lava flows, many vents, volcaniclastic deposits, and minor intrusions (Cameron and Stephenson, 1985). In the Bathgate area, a number of volcaniclastic deposits occur within the largely nonmarine, lower and middle Visean sedimentary sequence, while in upper Visean time, subaerial basaltic and basanitic lavas dominate the succession. The nature of the volcaniclastic rocks within the sequence will help to reconstruct the environment in which the various biota succumbed and were preserved.

Thin tuffs and lapilli tuffs occur throughout the limestone succession and may be of air-fall or epiclastic origin. An epiclastic origin requires erosion, redistribution, and deposition of volcanic fragments possibly from the flanks of a nearby volcano. In most volcanic environments, the quiescent periods between eruptions are dominated by epiclastic surface processes.

Other more violent processes may be ruled out because of the geotectonic setting and broadly basaltic nature of the volcanism. The Bathgate area must have been relatively close to a coastline in mid-Visean time, but Surtseyan volcanic activity, which has been reported elsewhere in the province, may be ruled out due to the lack of vesicularity in the volcaniclastic fragments.

Volcaniclastic facies analysis permits distinction between air-fall and epiclastic origins, since air-fall tuffs persist laterally. Preliminary bedform analysis suggests that individual tuff horizons vary greatly in thickness laterally. This may result from sediment being washed into existing hollows, suggesting an epiclastic origin.

Preliminary examination of clasts reveals rounding and variability in composition, which would support an epiclastic origin. It seems that fully articulated amphibian skeletons occur within the laminated limestones and that only disarticulated skeletons occur in the lapilli tuffs. This observation adds support to an epiclastic origin for the volcaniclastic rocks. Either the abundant, loose volcanogenic fragments were washed by equatorial rains (the abundance of wood suggests that the area was not arid) from the flanks of a small basaltic volcano or they moved by mass flow during a continuous eruptive episode onto or into an area of sinter deposition.

ACKNOWLEDGMENTS

Excavations at East Kirkton were generously permitted by West Lothian District Council, Stan Wood and W. R. Lawson, and carried out with funding from the Nature Conservancy Council. National Museums of Scotland staff, R. Reekie, W. Baird, C. Chaplin, and R. Paton are thanked for their work on site, which was supported by NMS. Fallick is supported by SURRC, funded by the Natural Environment Research Council and The Scottish Universities. A. Gunning and R. Sutcliffe of Glasgow City Museum and J. Campbell of Bennie Museum, Bathgate are thanked for collaborating during the excavations. The volume of collecting from the often-intractable limestone would not have been possible without the help of the following volunteers, who made a regular commitment of time to collect on site under NMS supervision: G. Baird, D. Beaumont, S. Brown, M. C. Drummond, M. Ford, M. Fothergill, J. Freeman, A. Henderson, W. Hill, M. T. Hutchison, A. Lee, F. Lindsay, S. Macdonald, R. Taylor, S. McLean, and N. Wyber.

A. Scott thanks J. Galtier for his collaboration and K. D'Souza for photography. Work with J. Galtier was supported by NATO grant 0036/87 and fieldwork by a grant from Royal Holloway and Bedford New College, University of London. Rolfe thanks The Royal Society of Edinburgh for a grant enabling Smithson's research collaboration. We thank J. Cater (Bristol University) for permission to quote the results of his vergence analysis, and M. Fothergill (Royal Botanic Gardens, Kew) for Figure 2.

REFERENCES CITED

Berger, B. R., and Bethke, P. M., 1985, Geology and geochemistry of epithermal systems: Society of Economic Geologists Reviews in Economic Geology, v. 2, 298 p.

Cadell, H. M., 1925, The rocks of West Lothian: Edinburgh, Oliver and Boyd, 390 p.

Cameron, I. B., and Stephenson, D., 1985, The Midland Valley of Scotland: London, British Regional Geology, HMSO, 172 p.

Cope, M. J., and Chaloner, W. G., 1985, Wildfire; An interaction of biological and physical processes, in Tiffney, B. H., ed., Geological factors and the evolution of plants: New Haven, Connecticut, Yale University Press, p. 257–277.

Dill, R. F., Shinn, E. A., Jones, A. T., Kelly, K., and Steinen, R. P., 1986, Giant subtidal stromatolites forming in normal salinity waters: London, Nature, v. 324, p. 55–58.

Fournier, R. O., 1985, The behaviour of silica in hydrothermal solutions, in Berger, B. R., and Bethke, P. M., eds., Geology and geochemistry of epithermal systems: Society of Economic Geologists Reviews in Economic Geology, v. 2, p. 45–62.

Francis, E. H., 1983, Carboniferous–Permian igneous rocks, in Craig, G. Y., ed., Geology of Scotland: Edinburgh, Scottish Academic Press, p. 297–342.

Galtier, J., 1986, Taxonomic problems due to preservation; Comparing compression and permineralized taxa, in Spicer, R. A., and Thomas, B. A., eds., Systematic and taxonomic approaches in palaeobotany: Oxford, Oxford University Press, p. 1–16.

Gee, H., 1988, Oldest known reptile found in Scotland: London, Nature, v. 336, p. 427.

Geikie, A., 1861, in Howell, H. H., and Geikie, A., The geology of the neighbourhood of Edinburgh: Memoirs of the Geological Survey of the United Kingdom, p. 1–151.

Hibbert, S., 1836, On the fresh-water limestone of Burdiehouse: Transactions of The Royal Society of Edinburgh, v. 13, p. 169–282.

Kjellesvig-Waering, E. N., 1986, A restudy of the fossil Scorpionida of the world: Palaeontographica Americana, v. 55, 287 p.

Large, D. J., 1987, The East Kirkton Limestone; A hot spring siliceous sinter [B.Sc. thesis]: Glasgow, University of Strathclyde, 22 p.

Milner, A. R., 1985, Scottish window on terrestrial life in the lower Carboniferous: London, Nature, v. 314, p. 320–321.

—— , 1988, The relationships and origin of living amphibians, in Benton, M. J., ed., The phylogeny and classification of the tetrapods: Oxford, Clarendon Press, v. 1, p. 59–102.

Milner, A. R., Smithson, T. R., Milner, A. C., Coates, M. I., and Rolfe, W.D.I., 1986, The search for early tetrapods: Modern Geology, v. 10, p. 1–28.

Monty, C.L.V., 1976, The origin and development of cryptalgal fabrics, in Walter, M. R., ed., Stromatolites; Developments in sedimentology: Amsterdam, Elsevier, v. 20, p. 193–249.

Muir, R. O., and Walton, E. K., 1957, The East Kirkton Limestone: Transactions of the Geological Society of Glasgow, v. 12, p. 157–168.

Oberst, C. H., 1987, Variation of Visean strata across the Midi thrust, Belgium [Ph.D. thesis]: University of Durham.

Panchen, A. L., and Smithson, T. R., 1988, The relationships of the earliest tetrapods, in Benton, M. J., ed., The phylogeny and classification of the tetrapods: Oxford, Clarendon Press, v. 1, p. 1–32.

Rex, G. M., and Scott, A. C., 1987, The sedimentology, palaeoecology, and preservation of the lower Carboniferous plant deposits at Pettycur, Fife, Scotland: Geological Magazine, v. 124, p. 43–66.

Rolfe, W.D.I., 1986a, Aspects of the Carboniferous terrestrial arthropod community; 9th International Congress on Carboniferous Stratigraphy and Geology, 1979, Washington, D.C., and University of Illinois at Urbana-Champaign: Compte Rendus, v. 5, p. 303–316.

—— , 1988, Early life on land; The East Kirkton discoveries: Earth Science Conservation, no. 25, p. 22–28.

—— , 1986b, Mr. Wood's fossils: Dinosaur Times, no. 12, p. 1–6.

Schopf, J. M., 1975, Modes of fossil plant preservation: Review of Palaeobotany and Palynology, v. 20, p. 27–53.

Scott, A. C., 1986, Distribution of lower Carboniferous floras in northern Britain; 9th International Congress on Carboniferous Stratigraphy and Geology, 1979, Washington, D.C., and University of Illinois at Urbana-Champaign: Compte Rendus, v. 5, p. 77–82.

Scott, A. C., and Rex, G. M., 1987, The accumulation and preservation of Dinantian plants from Scotland and its borders, in Miller, J., Adams, A. E., and Wright, V. P., eds., European Dinantian environments: Geological Journal Special Issue 12, p. 329–344.

Scott, A. C., Galtier, J., and Clayton, G., 1984, The distribution of lower Carboniferous anatomically preserved floras in western Europe: Transactions of the Royal Society of Edinburgh, Earth Sciences, v. 75, p. 311–340.

Scott, A. C., Meyer-Berthaud, B., Galtier, J., Rex, G. M., Brindley, S. A., and Clayton, G., 1986, Studies on a lower Carboniferous flora from Kingswood, near Pettycur, Scotland; 1, Preliminary report: Review of Palaeobotany and Palynology, v. 48, p. 161–180.

Selden, P. A., 1985, Eurypterid respiration: Philosophical Transactions of the Royal Society, series B, v. 309, p. 219–226.

Selden, P. A., and Jeram, 1989, Palaeophysiology of terrestrialization in the Chelicerata: Transactions of the Royal Society of Edinburgh, Earth Sciences, v. 80, p. 303–310.

Smithson, T. R., 1985, Scottish Carboniferous amphibian localities: Scottish Journal of Geology, v. 21, p. 123–142.

—— , 1989, The earliest known reptile: London, Nature (in press).

Steinen, R. P., Gray, N. H., and Mooney, J., 1987, A Mesozoic carbonate hot spring deposit in the Hartford Basin of Connecticut: Journal of Sedimentary Petrology, v. 57, p. 319–326.

Stephenson, D., Fortey, N. J., and Gallagher, M. J., 1983, Polymetallic mineralization in Carboniferous rocks at Hilderston, near Bathgate, central Scotland: Institute of Geological Sciences Mineral Reconnaissance Report 68, p. 1–42.

Waterston, C. D., Oelofsen, B. W., and Oosthuizen, R.D.F., 1985, *Cyrtoctenus wittebergensis* sp. nov. (Chelicerata: Eurypterida); A large sweep feeder from the Carboniferous of South Africa: Transactions of the Royal Society of Edinburgh, Earth Sciences, v. 76, p. 339–358.

Wellstead, C. F., 1982, A lower Carboniferous aïstopod amphibian from Scotland: Palaeontology, v. 25, p. 193–208.

Wood, S. P., Panchen, A. L., and Smithson, T. R., 1985, A terrestrial fauna from the Scottish lower Carboniferous: London, Nature, v. 314, p. 355–356.

MANUSCRIPT ACCEPTED BY THE SOCIETY JUNE 21, 1989

Preservation, evolution, and extinction of plants in Lower Carboniferous volcanic sequences in Scotland

Andrew C. Scott
Geology Department, Royal Holloway and Bedford New College, University of London, Egham, Surrey TW20 0EX, United Kingdom

ABSTRACT

Diverse anatomically preserved plant assemblages occur abundantly in Lower Carboniferous volcanic sequences in the Midland Valley Basin, Scotland. They are preserved as calcareous permineralizations and as fusain (fossil charcoal). The plants occur in basaltic ashes, lavas, peats, and limestones. Many of the assemblages occur in deposits that are interpreted as products of phreatomagmatic activity. Numerous genera and species occur specifically in volcanic rocks. The volcanic activity may have stimulated the diversification of early ferns and pteridosperms in particular, as well as causing ecological disturbance and vegetational change.

INTRODUCTION

The occurrence of plants in Tertiary to Recent volcanic rocks is widely known (e.g., Dorf, 1980; Fritz, 1980a, b; Karowe and Jefferson, 1987; Retallack, 1980). Even isolated fossil forests have been reported from early Mesozoic sequences (Jefferson, 1982). The occurrence of Paleozoic examples is less well known. It is not widely realized that plants in such environments are frequently anatomically (Schopf, 1975; Scott and Collinson, 1983; Scott, 1989) preserved. In some cases the plants occur as silica, carbonate permineralizations, or petrifactions (Schopf, 1975) or as fusain (fossil charcoal). The volcanic activity not only buries the plants, in some cases in situ, but also may be the cause of preservation. Volcanic environments are frequently unstable, and plants that colonize the mineral-rich soils may include ferns and horsetails. These plants successfully invaded volcanic environments during the early Carboniferous, and the availability of vacant habitats may have stimulated the diversification of some fern groups in particular (Scott and Galtier 1985; Collinson and Scott, 1987). Ferns and horsetails remain primary colonizers of such environments.

In the Lower Carboniferous of the Midland Valley of Scotland, volcanic rocks are widespread (Francis, 1983). They include plateau lavas, vent agglomerates, and basaltic ashes. Anatomically preserved plants occur widely in these rocks, mainly as calcareous permineralizations and fusain (Scott and Rex, 1987). The plants are well known and have been studied since the last century (e.g., Kidston, 1908; Scott, 1898; Gordon, 1935; Long, 1986; and reviewed in Scott and others, 1984). The diversity of plants and volcanic facies is notable, and preliminary observations have already been published (Scott and others, 1984, 1985, 1986; Scott and Rex 1987; Scott, 1990). In this chapter, I describe facies, geographic and stratigraphic distribution, diagenetic history, and ecological and evolutionary significance of the plants.

STRATIGRAPHIC AND GEOGRAPHIC DISTRIBUTION

Plant assemblages from the Lower Carboniferous (Mississippian) volcanics of the Midland Valley Basin occur at several stratigraphic levels and throughout the basin (Fig. 1). References to these localities are given in Appendix I. Late Tournaisian assemblages occur below the Clyde Plateau Lavas at Loch Humphrey Burn, north of Glasgow in the Kilpatrick Hills. In the east of the basin, the most diverse assemblages occur at Oxroad Bay in East Lothian. Mid-Visean assemblages occur only at Loch Humphrey Burn (and Glenarbuck?) in the west part of the basin. Late Visean assemblages associated with ashes and lavas occur at Pettycur, Kingswood, and Weaklaw, where in situ pteridosperms also are found. A probable late Visean in situ lycopod forest is known from Arran in the western part of the basin. The late Lower Carboniferous assemblage from East Kirkton is described in Durant and others (this volume).

Anatomically preserved plants are also common in nonvol-

Scott, A. C., 1990, Preservation, evolution, and extinction of plants in Lower Carboniferous volcanic sequences in Scotland, *in* Lockley, M. G., and Rice, A., eds., Volcanism and fossil biotas: Boulder, Colorado, Geological Society of America, Special Paper 244.

canic facies in the Midland Valley Basin and the Tweed Basin to the south (see details and references in Scott and others, 1984; Scott and Rex 1987; Scott in press). Several other localities yielding anatomically preserved plants in basaltic ashes have been discovered with J. Galtier and B. Meyer-Berthaud, but have yet to be described. Plant compression fossils are abundant in both volcanic and nonvolcanic facies, but will not be described here. A preliminary analysis of these compression floras has already been published (Scott, 1985).

VOLCANIC FACIES

Most of the plant assemblages occur in volcanic ashes. Representative lithologic logs of the main sequences are shown in Figure 2. Scott and Rex (1987) describe 5 facies associations in the Midland Valley Basin. More recent work at several localities, including Oxroad Bay, Pettycur, and Weaklaw, suggests a greater diversity of assemblages.

At the base of the Loch Humphrey Burn section (Fig. 2A) the plants occur mainly as fusainized fragments in basaltic ashes. Some calcareous permineralizations also occur. These ashes and agglomerates, which occur in an upward-fining sequence, include reworked lavas. The sequence has been dated by palynology as late Courceyean (late Tournaisian) in age. The plant assemblage comprises mainly fern and sphenopsid sporangia, but also includes leaf and stem fragments. The occurrence of such diverse early-fern sporangia has led to the suggestion that the diversification of these plants may be related to the spread of volcanic activity at this time and the ability of plants to colonize these new mineral-rich soils (Scott and Galtier, 1985). At Oxroad Bay there are several fossil plant assemblages. All occur in basaltic ashes. One is composed mainly of the lycopod *Oxroadia gracilis* Alvin. It occurs abundantly at several levels and dominates the assemblages. While it has been broken up and transported, the fragments are often current aligned, and occur with numerous cones. The plants and ecology of these beds are currently being investigated by R. M. Bateman. The ashes appear to be waterlain and vary in grain size. The only animals discovered are abundant scorpions, which currently are being studied by A. Jeram. Another plant assemblage, in coarser ashes within channels (Fig. 2b; Plate 1A, C) seen in the cliff section, contains the pteridosperm *Tetrastichia* together with the cupule *Salpingostoma* and the ovules *Tantallosperma* and *Stamnostoma* Plate 1E, F). These plants were the first to be described. They occur in calcareous lenses (Plate 1C), but layers of permineralized plants occur at several levels both below and above the main horizon. The plants include isolated axes (Plate 1F) and layers with ovules such as *Stamnostoma* (Plate 1E). The ashes vary considerably in grain size, but comprise mainly upward-fining units. One assemblage from the cliff section occurs within slumped blocks and yields large fragments of the pteridosperms *Buteoxylon* and *Triradioxylon* (Plate 1D). At this level, *Eristophyton* occurs as large permineralized logs. Toward the base of the cliff section (Fig. 2b), finer ashes commonly yield large pollen organs, which have yet to be described. Detailed studies of the plants and sedimentology of the cliff section is being undertaken with G. Rothwell and D. Wight. Clearly, the distribution of the plants at Oxroad Bay reflects original ecology, transport history, and diagenesis.

The mid-Visean assemblage at Loch Humphrey Burn (Fig. 2A) occurs in reworked ashes and comprises diverse pteridosperm foliage and fructifications preserved as partially permineralized compressions. The plants have been described by J. Walton and his students (see Scott and others, 1984), but new, recently discovered taxa will be described by R. M. Bateman. All of the plants have been transported, and the sediments may be interpreted as alluvially reworked volcanic rocks that underlie the main Clyde Plateau Lavas.

Late Visean plant assemblages occur in association with basalts and basaltic ashes. At Pettycur, 7 facies have been recently described (Rex and Scott, 1987). These include permineralized peats (Pettycur Limestone; Plate 2B) that yield an assemblage dominated by the lycopod *Paralycopodites* together with the fern *Botryopteris* and the pteridosperm *Heterangium*. Another peat contains the zygopterid ferns *Metaclepsydropsis* and *Diplolabis* (Plate 2C). Both of these limestones occur as rip-up blocks at the base of basalt lavas (Fig. 2f). In addition, ashes contain fragmented plants preserved as fusain and calcareous permineralizations. Occasionally the zygopterid peat was ripped up as layers, incorporated in the moving lava, and subsequently preserved by calcareous permineralization (Plate 2D). Some of the stems were also charred, probably by heat from the lava. It is possible that the peat may have been wet when incorporated; the lava is highly vesicular and broken. At the nearby locality of Kingswood, of the same age, anatomically preserved plants occur in limestones and ashy limestones above agglomerates (Fig. 2e; Plate 2E). The limestone contains the lycopod *Oxroadia* preserved as a calcareous permineralization, and pteridosperm axes, leaves, and pollen organs preserved as fusain (Plate 2F, G). New sections (Plate 2E) indicate a greater extent to this deposit than previously thought (Scott and others, 1986), and several plant-bearing horizons occur (Fig. 2e). In addition, fish scales and scorpion fragments are common. The relation of the two Kingswood sections and surrounding rocks is complex, complicated by faulting.

Late Visean plant assemblages also occur at Weaklaw (Fig. 1; Plate 3A–E). The sections are exposed only at low tide on the foreshore. The sequence includes typical oil shale facies that yield sparse plant compressions as well as several sequences of yellow dolomitic ashes. These ashes are folded and faulted, making the stratigraphic sequence difficult to unravel. Although anatomically preserved plants may be found at several levels along the shore, the most abundant and diverse assemblages occur within a syncline of ashes well exposed on the foreshore and described by Gordon (1927) and Day (1923). Paleosols are well developed at two levels and overlain by fine, and then coarse, ash (Fig. 2c, d; Plate 3A, B). The plants are preserved as compressions and calcareous permineralizations. In situ upright permineralized trunks of *Pitus* occur (see Gordon 1927, 1935; Plate 3C). In the fine-grained ashes overlying the paleosols, large fronds of *Spathulopte-*

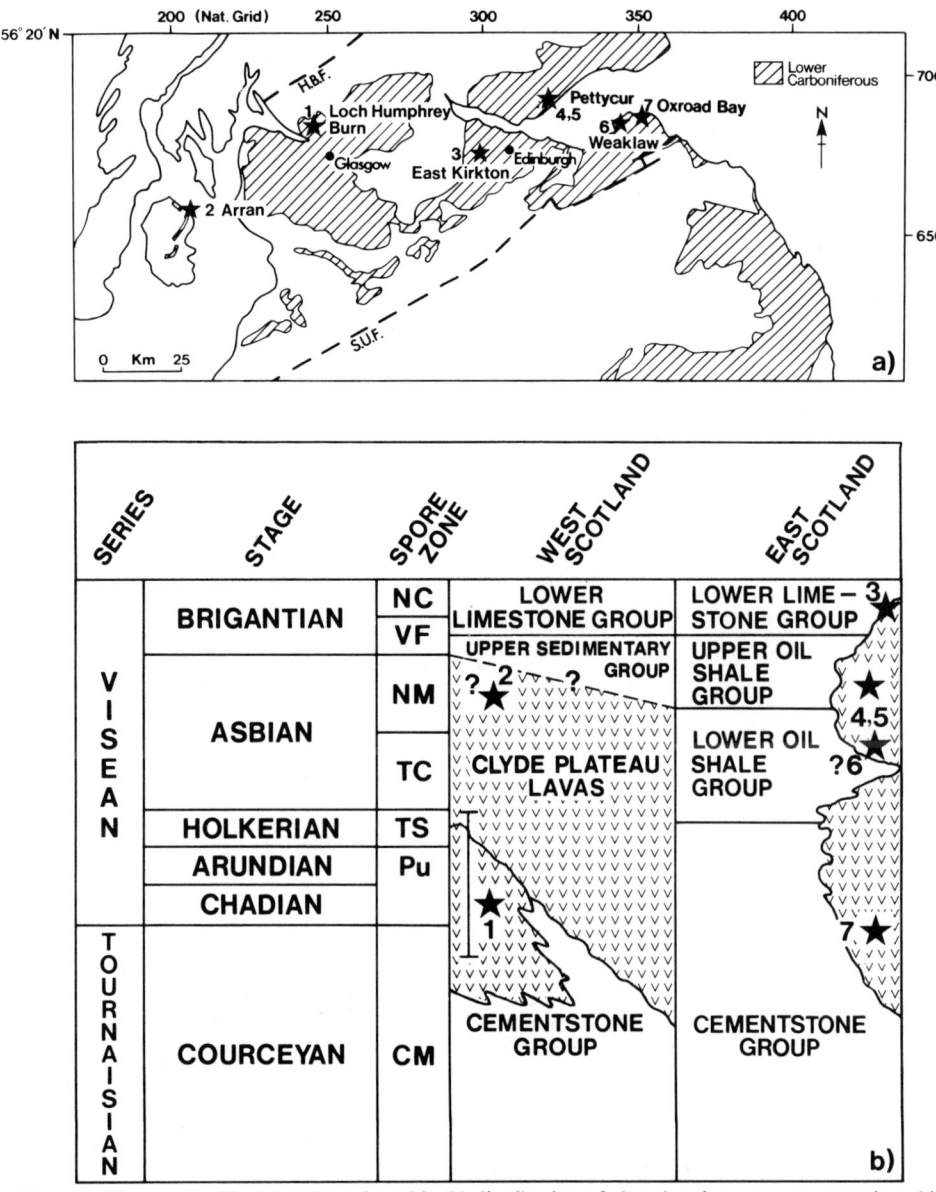

Figure 1. The geographic (a) and stratigraphic (b) distribution of plant-bearing sequences mentioned in the text. 1. Loch Humphrey Burn; 2. Arran; 3. East Kirkton; 4. Kingswood; 5. Pettycur; 6. Weaklaw; 7. Oxroad Bay.

ris and *Rhacopteris* occur (Plate 3D). In other parts of the sequence, isolated fragments of zygopterid ferns and lycopods occur. This locality lies 200 m to the west; the relation between the two outcrops is complicated by faulting and a dolerite intrusion.

In contrast, the in situ trunks of Arran occurring within basaltic ashes (Fig. 2g) are of the lycopod *Lepidophloios* (Plate 3F, G, H). Some of these are hollow and contain smaller plant fragments (Walton, 1935). Other plant remains are permineralized branches of *Lepidophloios* (Plate 3H), but some compressions have been found at the base of the sequence. The series of coarse green ashes and agglomerates with some shales overlies thick sandstones and is overlain by basalt lava flows. The dating of the sequence is uncertain, but it is probably of late Visean age.

The most recent discovery is from East Kirkton, where compressions are also partially preserved as calcareous permineralizations (see Durant and others, this volume). The plants are of late Visean (Brigantian) age and represent the youngest Dinantian anatomically preserved flora in the Midland Valley Basin. Partially permineralized compressions of several taxa of fern-like foliage occur in the limestones, including *Sphenopteridium*, *Sphenopteris*, and *Rhodea*. Several new taxa of anatomically preserved plants also occur. Both the overlying black shales and ashy sandstones contain plant compressions.

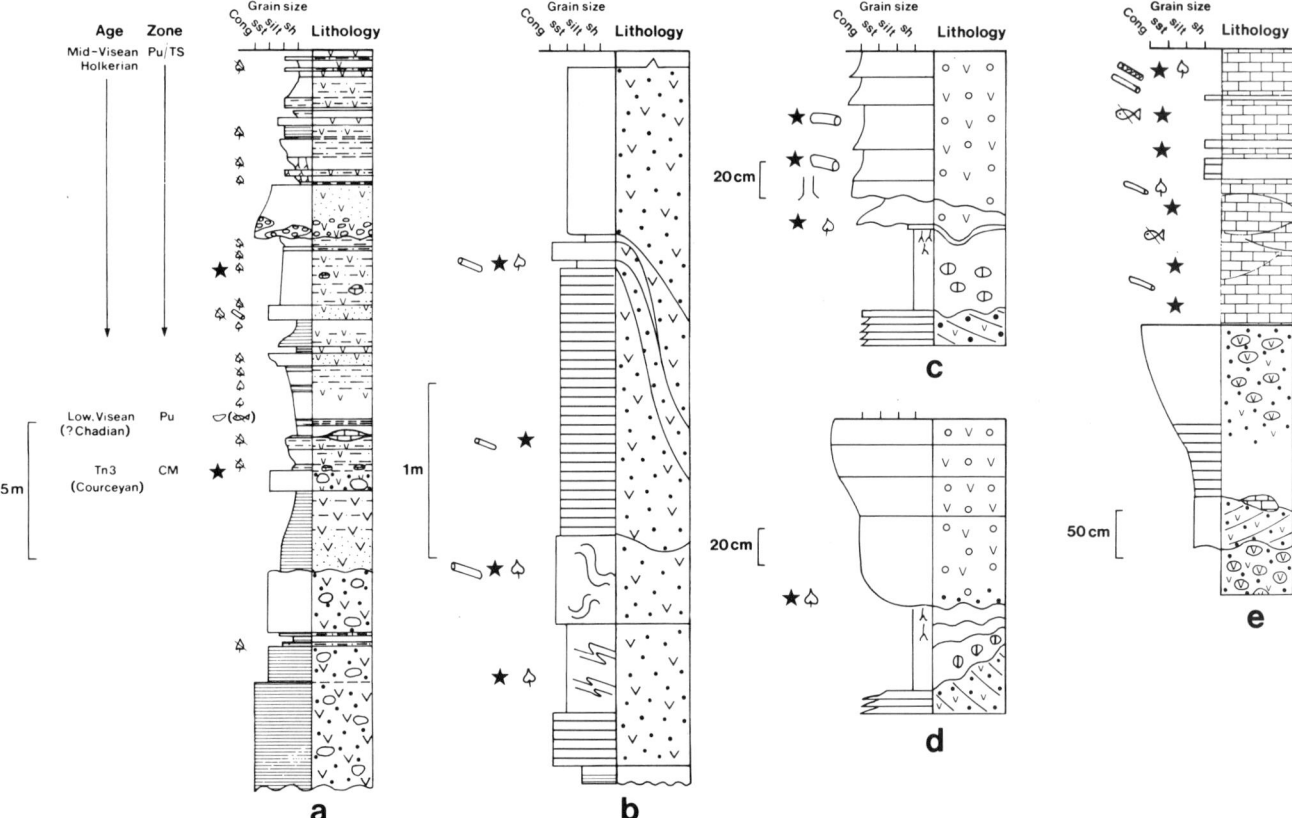

Figure 2. Lithological logs of the main sections mentioned in the text. (a) Loch Humphrey Burn; (b) Oxroad Bay cliff section; (c, d) Weaklaw rocks; (e) Kingswood new section; (f) Pettycur cliff section; (g) Arran shore section.

DIAGENESIS OF PLANTS

The permineralization of the plants vary considerably from locality to locality. The main features of the preservation are illustrated in Figure 3. In all cases the plants are anatomically preserved with carbonaceous cell walls remaining. The mode of permineralization varies considerably among localities and depositional setting. Preliminary studies have been made on several localities using thin sections, stained thin sections, and polished sections, the latter for cathodoluminescence and microprobe studies. Calcareous permineralizations from nonmarine clastic facies of Lower Carboniferous age all have a similar mineralization history. In these cases, there is an early calcite cement that often forms a rim to stems and a later ferroan calcite cement filling voids. There is little evidence of zoning in the calcite.

A Carboniferous lycopod branch found at Bearsden, near Glasgow, shows unusual fibrous crystals of calcite similar to those found in coal balls (Scott and Rex, 1985; Galtier and Scott, 1986; Fig. 3.4). The fibrous calcite has nucleated on cell walls and grown into voids or cell lumina. Undulating sharp boundaries to the crystal bundles are clearly seen. Marine influence appears to be an important factor in coal ball formation, and it may be the unusual mixing of marine and fresh water that stimulates calcite precipitation within the peat. At Bearsden, the environment appears to be alternating rapidly between marine and nonmarine, suggesting a similar mode of formation.

Permineralization in volcanic facies, however, is more diverse. Within permineralized peats, such as the Zygopterid and Pettycur Limestones from Pettycur, carbonate occurs as either large calcite crystals (Fig. 3.2) or as concentric cell fills (Fig. 3.1). In all cases, the carbonaceous cell walls remain. In many of the basaltic ashes, there is evidence of a complex cementation history, with plants being preserved in diverse ways. In some specimens, permineralization appears to have been rapid with a single cement generation, but in other specimens several generations of cement are evident. Occasionally layers of clay and calcite cover stems where water has continued to percolate through the sediment (Fig. 3.3). This is easily seen in specimens from Oxroad Bay. In all cases, calcite is abundant in the ashes and probably resulted from the breakdown of the basalts and basaltic ashes. Calcite is often found throughout the ashes, and filling cracks and vesicles in the lavas.

In some cases, the plants occur as fusain, which represent fossil charcoal, and is the result of wildfires (Cope and Chaloner, 1985) common in volcanic areas. Some of the charcoal may be the result of plants being charred by hot lava. The plants preserved in this manner may show excellent anatomy and are best studied using a scanning electron microscope.

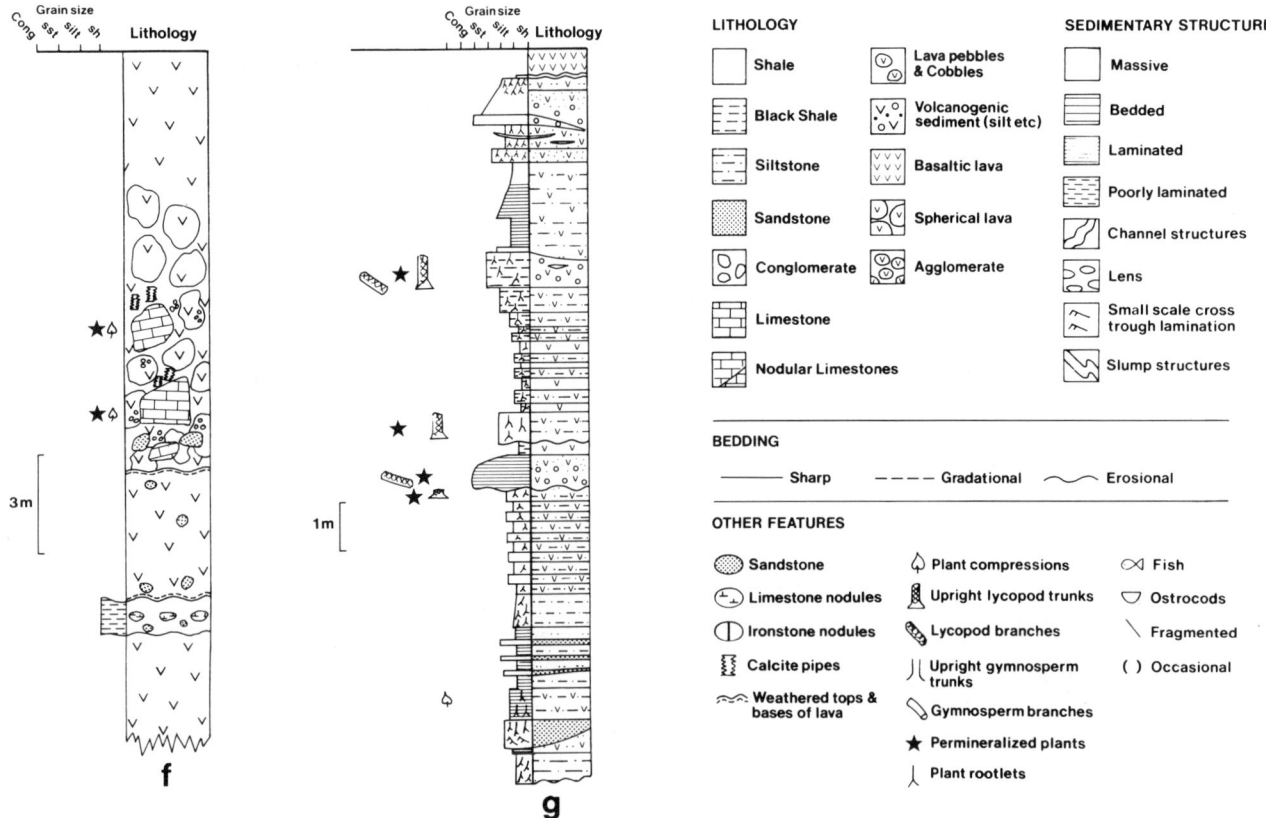

PALEOENVIRONMENTS AND PALEOECOLOGY

The Midland Valley volcanic sequences represent a diverse number of settings. These range from plateau lavas to vent deposits and ashes from ash cones formed by phreato-magmatic activity (Francis, 1983). Particularly significant are the diverse sequences of basaltic ashes and agglomerates from many localities throughout the basin and several stratigraphic levels. Many of the agglomerates occur within necks that have been interpreted as deeply dissected ash cones formed by explosive eruption as hot magma erupted through wet sediments (Francis, 1970; Leys, 1982). This type of phreatomagmatic activity gives rise to a wide range of volcanic form and product, including ash rings and maars (Lorenz, 1986; Kokelaar, 1986). Many of the features of bedding in the ashes, such as at Oxroad Bay, show some similarities with those described from modern maar volcanoes (Fisher and Waters, 1970).

At Pettycur, the volcanic sequences have been reconstructed (Rex and Scott, 1987) and show that the volcanoes erupted lavas and ashes (Fig. 4). Later, more lava eruption swamped the peats and ripped up blocks. At the margins of lakes, possibly crater lakes, at Kingswood, the margins were colonized by the lycopod *Oxroadia.*. Other plants were burned and washed into the lake.

The basaltic ashes were deposited in a variety of environments. At Oxroad Bay, laminated, rippled, and bedded ashes were deposited in a lake, possibly of a maar volcano (Lorenz, 1973). *Oxroadia* may have been a lakeside plant at this locality also. Large channels, which subsided rapidly during deposition, yield numerous plants. Some of these occur within slumped blocks and may have been living on the channel edges, whereas others were obviously transported some distance. Subsidence, slumping, and synsedimentary faulting are all common, as would be expected in a seismically active environment. The volcanoes may have covered a considerable area: it is difficult to assess how far some of the plants may have been transported. Fossil plant taxa abundant in the basalt ashes are usually rare elsewhere, and vice versa. Ferns and sphenopsids occur in the ashes at Loch Humphrey Burn. Here the plants are very fragmentary and may have traveled some distance. Paleo-environmental interpretation of this locality is problematic, but the occurrence of diverse early ferns may be significant and appears to be analogous to present-day settings of the type described by Spicer and others (1985) from Volcan Chichonal, Mexico. In all cases the volcanism may have caused severe disruption to the vegetation, giving rise to short- and long-term changes in plant communities. Distinguishing between pre- and post-eruptive vegetation may prove difficult.

EVOLUTION AND EXTINCTION

The Lower Carboniferous of Scotland has yielded a diversity of plants. Many well-known fossil-plant genera were first described from Scotland (see Scott and others, 1984, for review). In addition, numerous new taxa have been found in volcanogenic sediments during recent investigations. These include several fertile taxa. While individual beds may yield only a few species, the diversity of facies discovered has increased the number of new species encountered. Three groups of plants may be mentioned: ferns, lycopods, and pteridosperms.

Ferns

True ferns first appear in abundance in the Lower Carboniferous volcanogenic sediments of Loch Humphrey Burn (Galtier and Scott, 1985). They become more diverse in the later Visean, also occurring in volcanic settings. It is tempting to argue that their diversification at this time was stimulated by volcanism, especially as ferns are often pioneer plants on Recent lava and ashes (e.g., Spicer and others, 1985). Some zygopterid ferns, e.g., *Diplolabis* and *Metaclepsydropsis* with their scrampling habit, appear well suited for colonizing unstable ashes (Scott and Galtier, 1985). However, it is difficult to know whether this view is influenced by the fact that the plants may be preferentially preserved in volcanic settings. Several fern genera did not survive the end of the Lower Carboniferous, but again the reason for extinction is unclear. Ferns, as a whole, continue to diversify during the Upper Carboniferous (Galtier and Scott, 1985). It is interesting to note that spore-bearing plants have remained important pioneers in volcanic terrains (Spicer and others, 1985), so that, while the taxa have changed, the basic ecological structure has not (Collinson and Scott, 1987).

Lycopods

The lycopods are more difficult to assess. In the late Tournaisian, fluvio-lacustrine clastic sediments of Scotland yield abundant *Lepidodendron calamopsoides*. Its megaspore *Lagenicula crassaculeata* is ubiquitous in the sediments. When volcanic ash entered the sequence, this plant could not survive and is not found in volcanogenic sediments. In contrast the smaller "herbaceous" lycopod *Oxroadia* occurs abundantly in ashes, but rarely elsewhere. In the late Tournaisian of Oxroad Bay and late Visean of Kingswood, *Oxroadia* appears to be preserved in lake-margin settings, where there is significant ash input, perhaps indicating that the plant was able to survive in unstable conditions. The arborescent lycopod *Lepidophloios wünchianus* is also found in coarse, water-lain, basaltic ashes in the late Visean of Arran. There are numerous rootlet horizons throughout the sequence with no aerial portions of the plants preserved, but 0.5 to 1 m coarse ashes appear to have inundated and preserved several trunks at least on three occasions. Each trunk is no more than 1 m

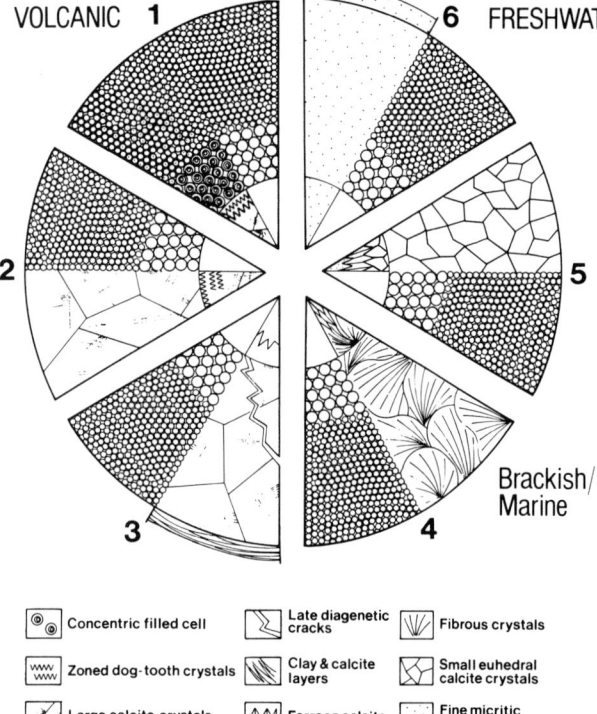

Figure 3. Calcareous permineralization of Scottish Carboniferous plants. Each segment is divided into two. In one half the cellular structure of the plant is shown diagrammatically with a central void and two types of other cells. In the other half these cells have generally been omitted to show the mineralization. (1-3) Mineralization in volcanic terrains. (1) Permineralization of a stem from the Pettycur lava. Here the cells have been concentrically filled with calcite. The zoning is easily seen in cathodoluminescence. (2) Permineralization of a stem from the Pettycur Limestone. Note difference from 1. (3) Permineralization of a stem from the cliff section at Oxroad Bay. There are several cement generations. Late diagenetic cracks are common. Some stems show an outer layer composed of calcite and clays. (4) Permineralization of a stem from early Namurian, brackish marine shales near Glasgow. This style of permineralization is similar to that found in coal balls. (5) Permineralization of a stem from late Tournaisian, freshwater, shelly limestones from Berwickshire. Note late ferroan calcite in the void fills. (6) Permineralization of a stem from fluvio-lacustrine sediments of late Tournaisian age in Berwickshire. The cement in the stem and of the rim are of the same generation.

Figure 4. Hypothetical reconstruction of the original paleoecology, deposition, and fossilization of the plant deposits at Pettycur and Kingswood (modified from Rex and Scott, 1987). (a) Original paleoecology showing growth and fossilization sites. The Pettycur Limestone flora (1) established a peat-forming area on the volcanic surface during a period of quiescence in the volcanic activity. The zygopterid ferns (2) established small-scale peat areas, which were frequently burned and the plants transported into small lakes (3) where carbonate was being precipitated. Pteridosperms grew in a separate site (4) and were transported as debris into small lakes on the volcanic surface in which dolomitic mudstones were being deposited (5). The Kingswood flora was separated by time or a barrier as shown here from the Pettycur flora. In this area pteridosperms and gymnosperms grew at some distance from the lake, and were frequently burned and transported into the lake as fusain (6). Nearer the lake, *Oxroadia* (7) grew, was washed into the lake, and permineralized in the limestone that was being deposited. (b) Subsequent period of eruption of nearby volcanic centers. This results in a series of ash falls and lava flows that destroy the peat-forming areas (8) and the zygopterid limestones (9). A nearby eruption modifies the Kingswood Limestone (10), resulting in local brecciation and incorporation in a pyroclastic series. (c) Final burial and relative positions of the plant-bearing deposits at the present day. The true peat lithology (Pettycur Limestone) (11) and the Zygopterid Limestone and ashy limestones (12) are incorporated into the base of basaltic lavas, which flowed over the lithified deposits. Plant compression fossils occur within dolomitic mudstones (13), which outcrop between lava flows. The Kingswood Limestone (14) occurs as blocks and as bedded limestones within an agglomerate sequence, which may represent a down-faulted block formed as a result of subsidence in the volcanic terrain.

Plants in Lower Carboniferous volcanic sequences 31

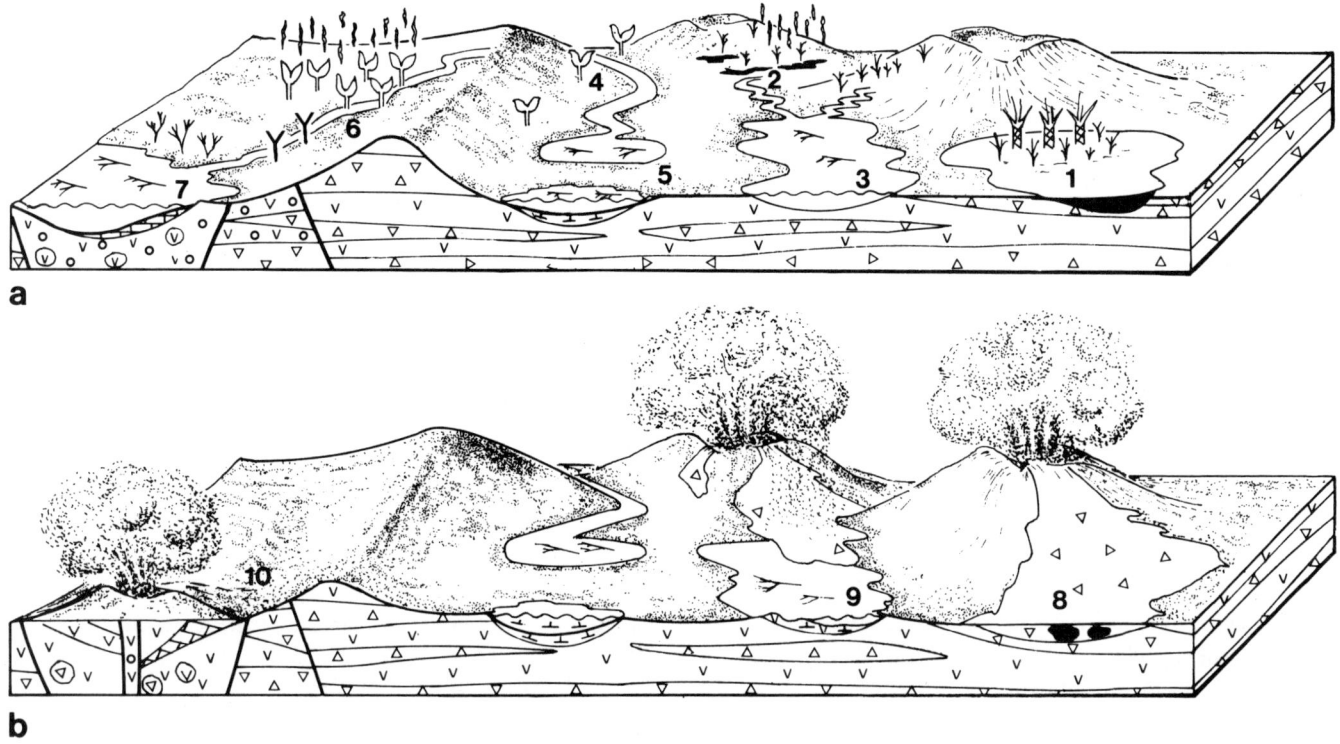

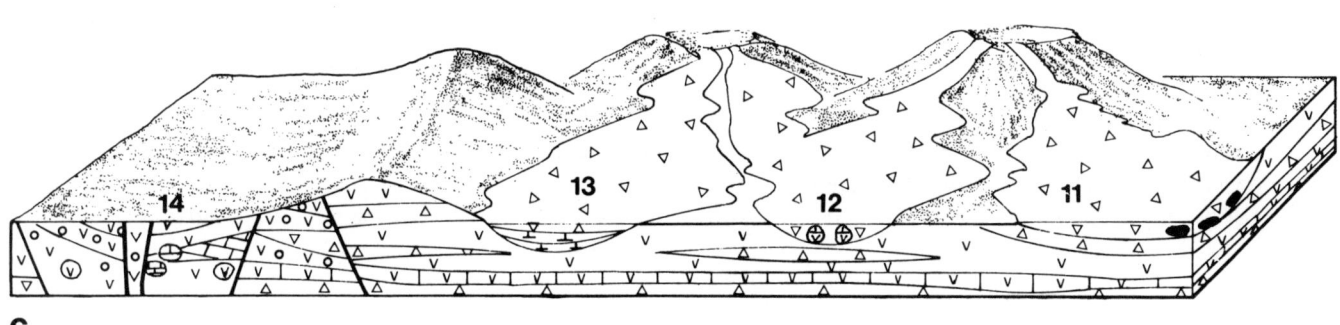

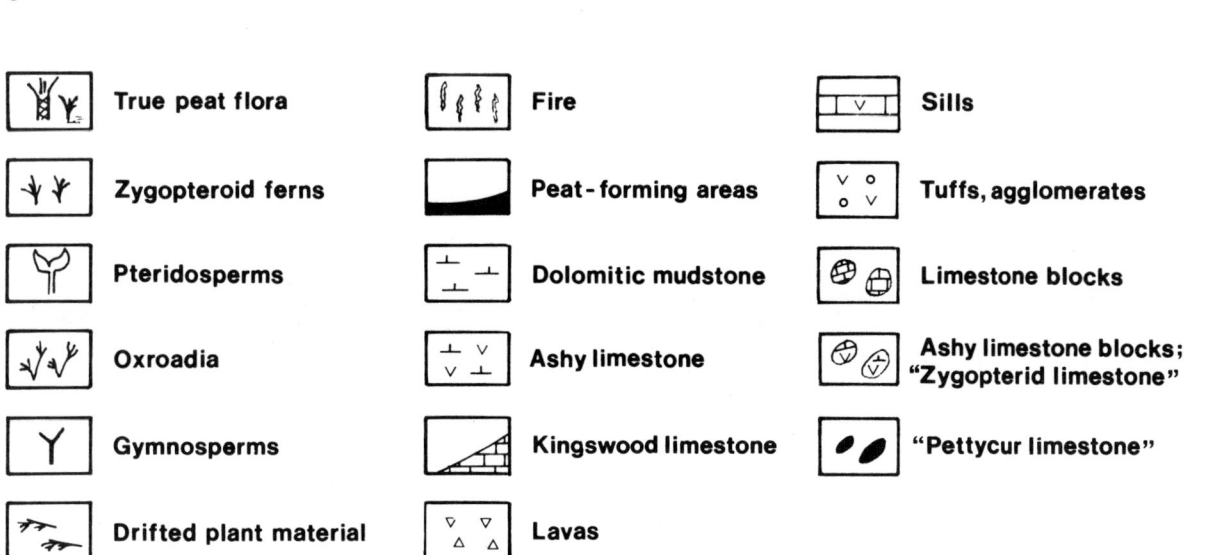

True peat flora	Fire	Sills
Zygopteroid ferns	Peat-forming areas	Tuffs, agglomerates
Pteridosperms	Dolomitic mudstone	Limestone blocks
Oxroadia	Ashy limestone	Ashy limestone blocks; "Zygopterid limestone"
Gymnosperms	Kingswood limestone	"Pettycur limestone"
Drifted plant material	Lavas	

high and has been cut off by the overlying ash (Plate 3F). There may have been considerable time between units as some trunks are hollow and plants have been washed in (Walton, 1935). The arborescent lycopod *Paralycopodites brevifolius* is known throughout the Carboniferous, and although a major contributor to the Pettycur peat, it is not specifically associated with volcanic environments. *Lepidodendron, Lepidophloios,* and *Paralycopodites* continued into the Upper Carboniferous where they became dominant elements of the flora. At this time they appear to have been well adapted to wet stable habitats and only became extinct because of major climatic changes (DiMichele and others, 1987). The herbaceous *Oxroadia* is as yet unknown in the Upper Carboniferous and may have become extinct at the end of the Lower Carboniferous, perhaps because of a change in the style of volcanism.

Pteridosperms

The Lower Carboniferous represents a major phase of seed-plant evolution. A wide diversity of seeds have been described by A. G. Long from the fluvio-lacustrine Cementstone Group (see Scott and others, 1984, for a review). Several other taxa occur in the late Tournaisian ashes at Oxroad Bay. While there are taxa in common between the two floras, some are exclusive to the volcanic environments. It is possible that the variety of environments, both volcanic and nonvolcanic, encouraged a diversity of these plants, with a number of experimental lines spreading into vacant ecological niches. The diversity of ovule types apparently decreased during the Upper Carboniferous. Likewise, pteridosperm generic diversity also appears to have decreased, but those taxa that survived dominated non-peat-forming environments. Pollen organ diversity, in contrast, increases throughout the Upper Carboniferous, and several distinct lines are recognizable (Meyer-Berthaud, 1987). Many pollen organs appear in Lower Carboniferous volcanogenic rocks for the first time. For example, two new pollen organs, *Melissiotheca* and *Phacelotheca,* have been described only from Kingswood (Meyer-Berthaud, 1986; Meyer-Berthaud and Galtier, 1986).

CONCLUSIONS

Anatomically preserved plants occur abundantly in Lower Carboniferous volcanic sequences in the Midland Valley Basin, Scotland. Diverse plant assemblages occur in various volcanic settings, associated with lavas and ashes, especially from phreatomagmatic volcanoes. Preservation is by calcareous permineralization or as fusain. The carbonate for plant permineralization probably came from the breakdown of basaltic ashes and lavas. The mechanisms of permineralization are diverse.

Some plants appear to be specifically associated with volcanic activity, and some plants may have colonized virgin mineral-rich soils. The relation between extinctions and volcanic activity is less clear.

Volcanic sequences are an important source of plant fossils that, to date, have not been studied systematically with a view to understanding the circumstances under which the floras were preserved. Further study promises to elucidate the paleoecology and taphonomy, evolution and extinction of terrestrial vegetation in these environments.

ACKNOWLEDGMENTS

I thank J. Galtier, B. Meyer-Berthaud, and G. Rex for discussion and field assistance, and C. Hildrew for drafting the diagrams, K. de Souza for photography, and S. Viggers for typing the manuscript. This work was supported by NERC grant GR3/4986, NATO grants RG361/83 and 0036/87, and a grant from the University of London Central Research Fund.

PLATE 1

Fossil plants from volcanogenic sediments of late Tournaisian age, Oxroad Bay, East Lothian, Scotland.

A. Cliff section showing bedded ashes within channel structure caused by subsidence during deposition. Position of Fig. C arrowed.

B. Thinly bedded ashes yielding plant compressions and some permineralizations from the foreshore.

C. Calcareous lense from coarse ashes of the cliff yielding anatomically preserved plants.

D. Slumped block yielding large permineralized frond of *Buteoxylon* and *Triradioxylon.*

E. Thin section of ash from cliff, yielding numerous ovules of *Stamnostoma* preserved as calcareous permineralization ×4 (OX397).

F. Pteridosperm axes from coarse ash from cliff section preserved as calcareous permineralization ×2½ (OX402).

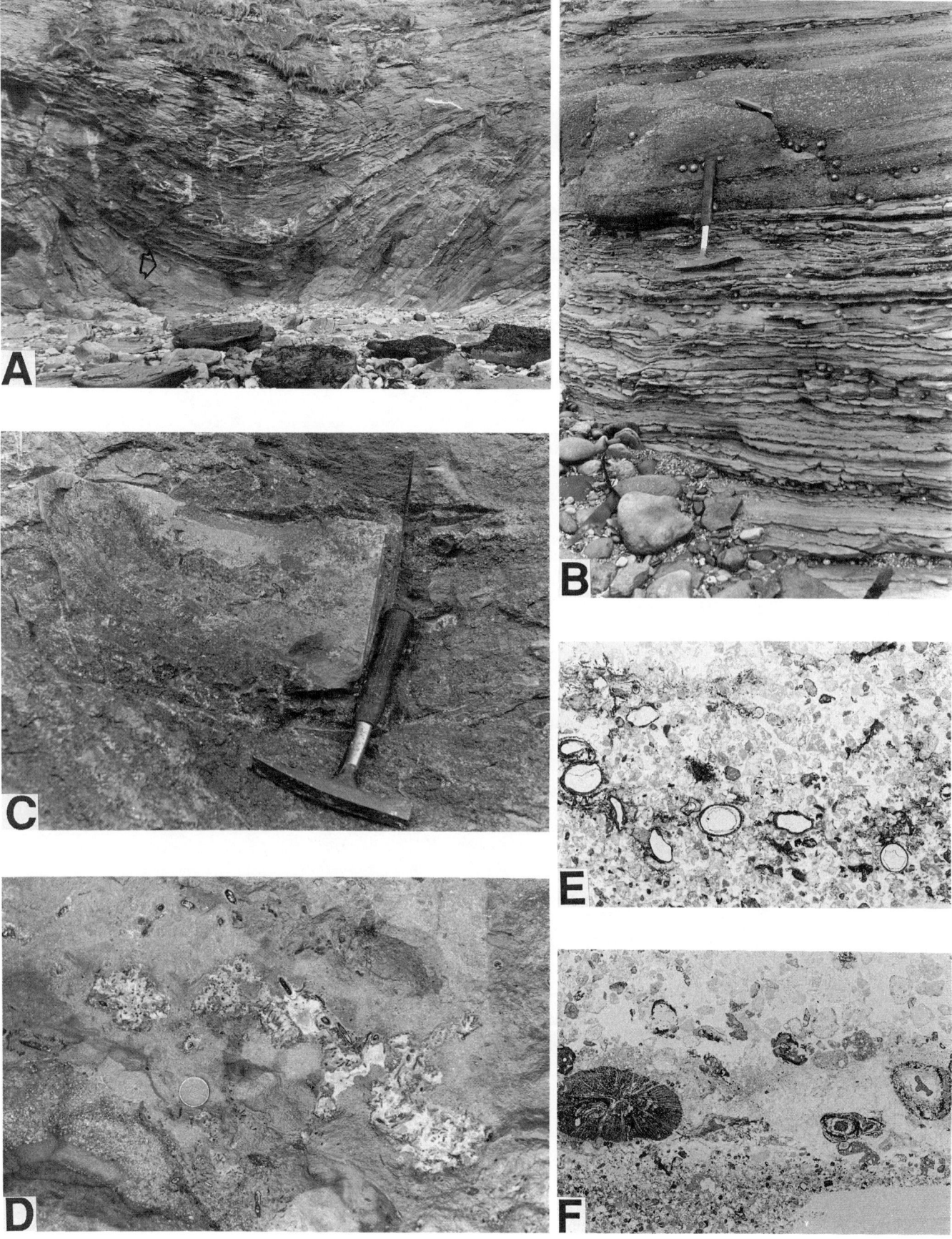

PLATE 2

Fossil plants from lavas and volcanogenic sediments of late Visean age, Pettycur (A–D) and Kingswood (E–G), Fife, Scotland.

A. Permineralized axis of zygopterid fern from lava at Pettycur.

B. Thin section of Pettycur Limestone dominated by lycopods and ferns ($\times 3$).

C. Zygopterid limestone with layers of fusainized and calcareous permineralizations *Diplolabis* and *Metacelpsydropsis* ($\times \frac{1}{2}$).

D. Pettycur Harbour peat within lava. Note zygopterid fern (arrowed).

E. New section (1987) at Kingswood showing bedded agglomerates and ashes (A) overlain by plant-bearing limestones (L).

F. Kingswood Limestone with fusainized wood (W) and pteridosperm axes (P) with permineralized lycopods (L) including *Achlamydocarpon* ($\times 3$).

G. Kingswood Limestone with fusainized plant fragments including pollen organs (P) and plant-bearing coprolites (C) ($\times 4$).

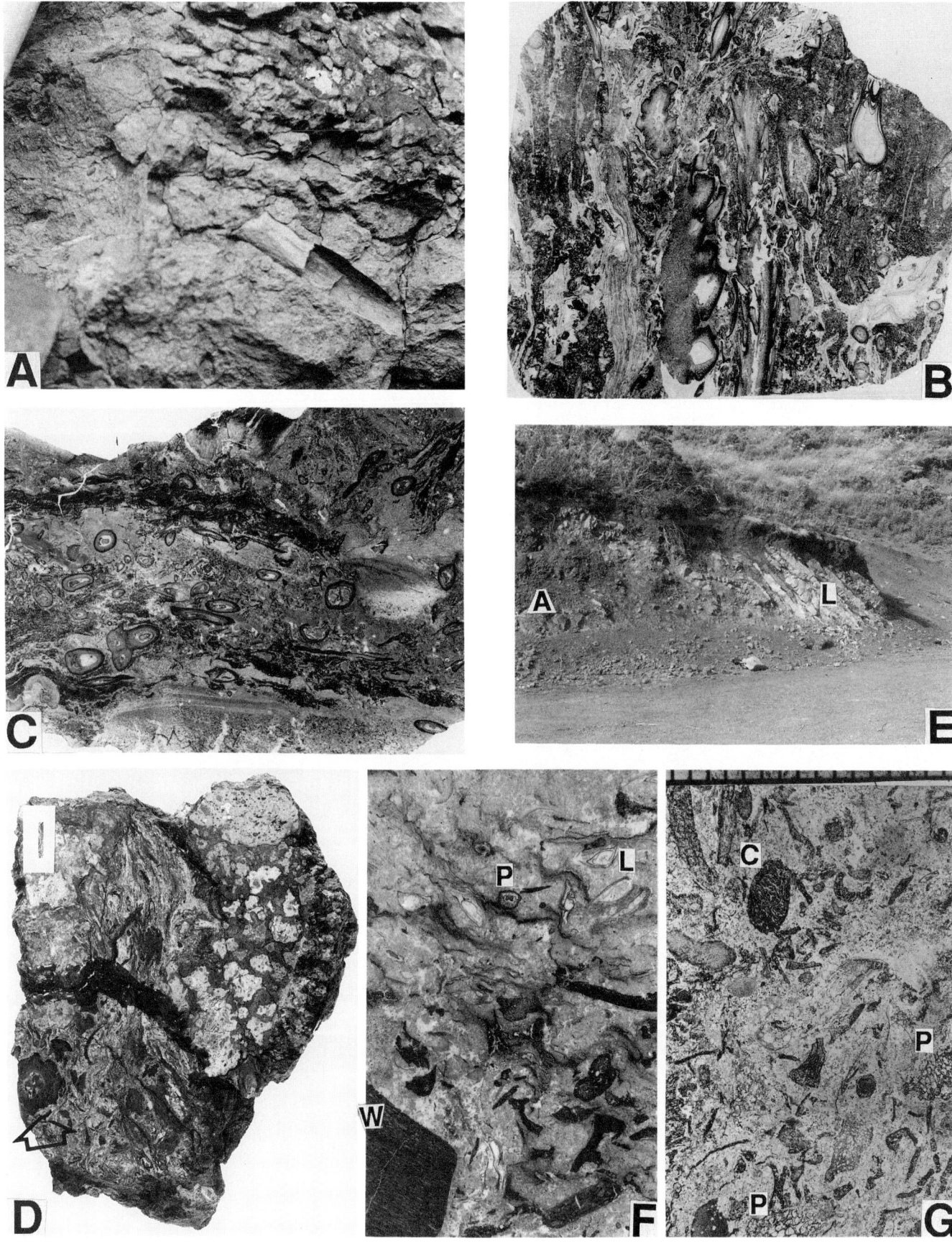

PLATE 3

Fossil plants from volcanogenic sediments of late Visean age from Weaklaw rocks, North Berwick, East Lothian (A–E) and from Arran (F–H).

A. Eastern plant bed, Weaklaw Rocks showing paleosol (P), compression horizon (C), and overlying dolomitic ash (A).

B. Western plant bed, Weaklaw Rocks showing underlying ashes, paleosol (P), and plant-bearing dolomitic ashes (A).

C. Upright permineralized *Pitus* trunk viewed from above, eastern plant bed, Weaklaw rocks.

D. Plant compressions of *Spathulopteris* and *Rhacopteris* from eastern plant bed (scale 1 cm).

E. Section through ash and compression horizon showing partial permineralization of the plants. Section in correct orientation. Scale 1 cm.

F. Upright trunk of *Lepidophloios* from basaltic ashes, Arran.

G. *Stigmaria* of *Lepidophloios* from basaltic ashes, Arran.

H. Section of basaltic ash with branch of *Lepidophloios* preserved as calcareous permineralization, Arran (scale 1 cm).

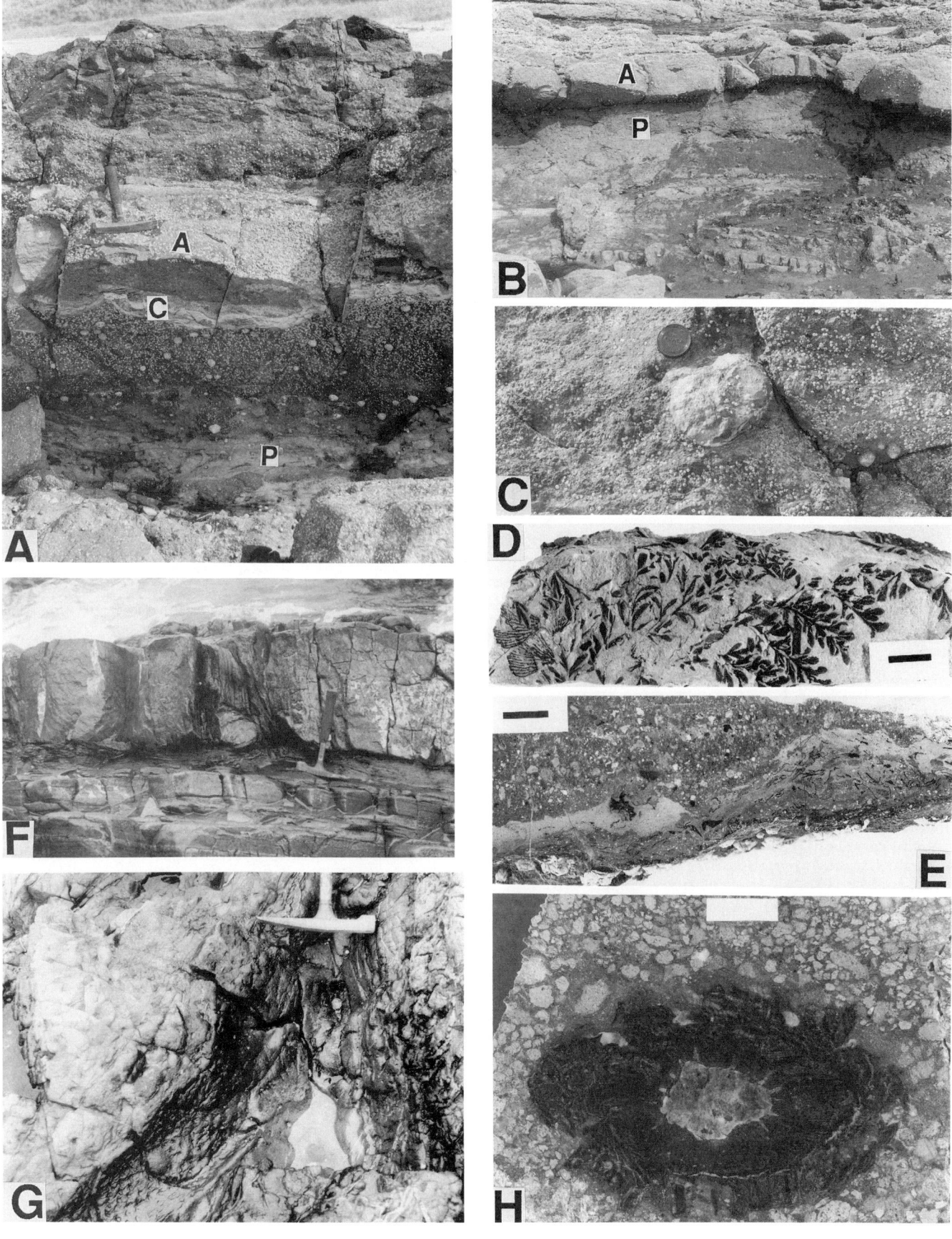

Appendix I: Localities mentioned in the text

1. Loch Humphrey Burn, late Tournaisian–Visean basaltic ashes (Scott and others, 1984, 1985).
2. Arran, late Visean basaltic ashes (Walton, 1935).
3. East Kirkton, late Visean limestones and shales (Durant and others, this volume).
4. Kingswood, late Visean limestone within basaltic ashes (Scott and others, 1986).
5. Pettycur, late Visean limestones and basaltic ashes (Rex and Scott, 1987).
6. Weaklaw rocks, late Visean dolomitic ashes (Gordon, 1927, 1935).
7. Oxroad Bay, late Tournaisian basaltic ashes (Scott and others, 1984, and including foreshore exposures).

REFERENCES CITED

Collinson, M. E., and Scott, A. C., 1987, Factors controlling the organization and evolution of ancient plant communities, in Gee, J.H.R., and Giller, P. S., eds., Organization of communities: Oxford, Blackwell Scientific Publication, p. 399–420.

Cope, M., and Chaloner, W. G., 1985, Wildfire; An interaction of biological and physical processes, in Tiffney, B. H., ed., Geological factors and the evolution of plants: New Haven, Connecticut, Yale University Press, p. 257–277.

Day, T. C., 1923, A new volcanic vent and other new geological features on the shore, Weaklaw, near Gullane: Transactions of the Edinburgh Geological Society, v. 11, p. 187–192.

Dorf, E., 1980, Petrified forests of Yellowstone: Washington, D.C., U.S. Government Printing Office, 31 p.

DiMichele, W. A., Phillips, T. L., and Olmstead, R. G., 1987, Opportunistic evolution; Abiotic environmental stress and the fossil record of plants: Review of Palaeobotany and Palynology, v. 50, p. 151–178.

Fisher, R. V., and Waters, A. C., 1970, Base surge bedforms in maar volcanoes: American Journal of Science, v. 268, p. 157–180.

Francis, E. H., 1970, Bedding in Scottish (Fifeshire) tuffpipes and its relevance to maars and calderas: Bulletin Volcanologique, v. 34, p. 697–712.

—— , 1983, Carboniferous-Permian igneous rocks, in Craig, G. Y., ed., Geology of Scotland, 2nd ed., Edinburgh, Scottish Academic Press, p. 297–324.

Fritz, W. J., 1980a, Reinterpretation of the depositional environment of the Yellowstone "fossil forests": Geology, v. 8, p. 309–313.

—— , 1980b, Stumps transported and deposited upright by Mount Saint Helens ash flows: Geology, v. 8, p. 586–588.

Galtier, J., and Scott, A. C., 1985, The diversification of early ferns: Proceedings of the Royal Society of Edinburgh, series B, v. 86, p. 289–301.

—— , 1986, A partially permineralized *Lepidophloios* from the early Upper Carboniferous of Scotland: Annals of Botany, v. 58, p. 617–626.

Gordon, W. T., 1927, The coastal strip between North Berwick and Cheese Bay, Gullane: Proceedings of the Geologists Association, v. 38, p. 441–446.

—— , 1935, The genus *Pitys* Witham emend: Transactions of the Royal Society of Edinburgh, v. 58, p. 279–311.

Jefferson, T. H., 1982, Fossil forests from the Lower Cretaceous of Alexander Island, Antarctica: Palaeontology, v. 25, p. 1–47.

Karowe, A. L., and Jefferson, T. H., 1987, Burial of trees by eruptions of Mount St. Helens, Washington; Implications for the interpretation of fossil forests: Geological Magazine, v. 124, p. 191–204.

Kidston, R., 1908, On a new species of *Dineuron* and of *Botryopteris* from Pettycur, Fife: Transactions of the Royal Society of Edinburgh, v. 46, p. 361–364.

Kokelaar, P., 1986, Magma-water interactions in subaqueous and emergent basaltic volcanism: Bulletin of Volcanology, v. 48, p. 275–290.

Leys, C., 1982, Volcanic and sedimentary processes in phreatomagmatic volcanoes [Ph.D. thesis]: University of Leeds.

Long, A. G., 1986, Observations on the Lower Carboniferous lycopod *Oxroadia gracilis* Alvin: Transactions of the Royal Society of Edinburgh, Earth Sciences, v. 77, p. 127–142.

Lorenz, V., 1973, On the formation of maars: Bulletin Volcanologique, v. 37, p. 183–204.

—— , 1986, On the growth of maars and diatremes and its relevance to the formation of tuff rings: Bulletin of Volcanology, v. 48, p. 265–274.

Meyer-Berthaud, B., 1986, *Melissiotheca* gen. nov. a pollen organ from the Upper Visean of Scotland: Botanical Journal of the Linnean Society, v. 93, p. 270–290.

—— , 1987, Contribution a l'étude des gymnospermes du Carbonifère inferieur D'Europe [Thèse d'Etat]: Montpellier, France, Université des Sciences et Techniques du Languedoc, 131 p.

Meyer-Berthaud, B., and Galtier, J., 1986, Studies on a new Lower Carboniferous flora from Kingswood, near Pettycur, Scotland; 2, *Phacelotheca*–a new synangiate fructification of pteridospermous affinity: Review of Palaeobotany and Palynology, v. 48, p. 180–198.

Retallack, G., 1980, Comment on "Reinterpretation of the depositional environment of Yellowstone 'fossil forests'": Geology, v. 9, p. 52–53.

Rex, G. M., and Scott, A. C., 1987, The sedimentology, palaeoecology, and preservation of the Lower Carboniferous plant deposits at Pettycur, Fife, Scotland: Geological Magazine, v. 124, p. 43–66.

Schopf, J. M., 1975, Modes of fossil preservation: Review of Palaeobotany and Palynology, v. 20, p. 27–53.

Scott, A. C., 1985, Distribution of Lower Carboniferous Floras in northern Britain: IX International Congress on Carboniferous Stratigraphy and Geology, Illinois 1979: Compte Rendu, v. 5, p. 77–82.

Scott, A. C., 1989, Anatomical preservation of fossil plants, in Briggs, D.E.G., and Crowther, P., eds., Encyclopaedia of palaeobiology: Oxford, Blackwell Scientific Publications, p. 263–266.

—— , 1990, The ecology of Scottish Lower Carboniferous floras; Preliminary observations; 11th Congrès International de Stratigraphie et du Geologie du Carbonifère, Beijing 1987: Compte Rendu (in press).

Scott, A. C., and Collinson, M. E., 1983, Investigating fossil plant beds; 1, The origin of fossil plants and their sediments: Geology Teaching, v. 7, p. 114–122.

Scott, A. C., and Galtier, J., 1985, The distribution and ecology of early ferns: Proceedings of the Royal Society of Edinburgh, series B, v. 86, p. 141–149.

Scott, A. C., and Rex, G. M., 1985, The formation and significance of Carboniferous coal balls: Philosophical Transactions of the Royal Society, series B, v. 311, p. 123–139.

—— , 1987, The accumulation and preservation of Dinantian plants from Scotland and its borders, in Miller, J., Adams, A. E., and Wright, V. P., eds., European Dinantian environments: Chichester, J. Wiley and Sons, p. 329–344.

Scott, A. C., Galtier, J., and Clayton, G., 1984, The distribution of Lower Carboniferous anatomically preserved floras in Western Europe: Transactions of the Royal Society of Edinburgh, Earth Sciences, v. 75, p. 311–340.

—— , 1985, A new late Tournaisian (Lower Carboniferous) flora from the Kilpatrick Hills, Scotland: Review of Palaeobotany and Palynology, v. 44, p. 81–99.

Scott, A. C., Meyer-Berthaud, B., Galtier, J., Rex, G., Brindley, S. A., and Clayton, G., 1986, Studies on a new Lower Carboniferous flora from Kingswood near Pettycur, Scotland; 1, Preliminary report: Review of Palaeobotany and Palynology, v. 48, p. 161–180.

Scott, D. H., 1898, On the structure and affinities of fossil plants from the Palaeozoic rocks; Part 1, On *Cheirostrobus*: Philosophical Transactions of the Royal Society of London, series B, v. 189, p. 1–34.

Spicer, R. A., Burnham, R. J., Grant, P., and Glicken, H., 1985, *Pityogramma calomelanos,* the primary, post-eruption colonizer of Volcan Chichonal, Chiapes, Mexico: American Fern Journal, v. 75, p. 1–5.

Walton, J., 1935, Scottish lower Carboniferous plants; The fossil hollow trees of Arran and their branches (*Lepidophloios wünchianus* Carruthers): Transactions of the Royal Society of Edinburgh, v. 58, p. 313–337.

Manuscript Accepted by the Society June 21, 1989

Printed in U.S.A.

The role of volcanism in K/T extinctions

Alan Rice
Department of Physics, University of Colorado at Denver, 1200 Larimer Street, Denver, Colorado 80204

ABSTRACT

Worldwide volcanism coincident with the end of the Cretaceous provides the source of iridium, shocked minerals, the carbon (CO_2) excursion, and the attendant greenhouse effect (from ^{18}O data). Reproductive stress under greenhouse conditions may have been a deciding factor in extinction of terrestrial fauna. Acidification of the upper ocean due to volcanic CO_2 may be the responsible factor in the K/T nannoplankton event. The K/T transition is characterized by extensive paludification, forming bryophytic bogs, the source of K/T coal beds. Mantle convection is deemed the source of the worldwide volcanism attending the Late Cretaceous. Episodic mantle convection may have been responsible for other extinction events.

THE K/T DEBATE

In 1980, Alvarez and others proposed that the great extinctions demarking the K/T transition were brought about by a massive meteor impact. The impact was to have lofted so much debris into the Earth's atmosphere as to create a "nuclear winter," which caused much of the life on Earth to perish. This arose from the discovery of an iridium anomaly associated with sedimentary rock that lay at the K/T transition. The iridium content was far greater than the average world abundance, suggesting a meteoritic source. Since the impact proposal, however, anomalously large amounts of iridium were discovered to be spewing forth from the Hawaiian volcanoes (Olmez, and others, 1986; Zoller and others, 1983), suggesting that the events at the K/T boundary could have had a volcanic origin (e.g., Officer and others, 1987). Even a rather small, dimunitive volcanic event (in comparison to the total geologic record) can cause mass extinctions. A particularly interesting example is the Laki fissure eruption in Iceland in 1783. Only 12 km^3 of lava erupted, and this was spread over an eight-month period. Twenty-four percent of the Icelanders perished, along with 75 percent of the livestock. Although it is estimated that the eruption dumped a billion tons of sulfuric acid into the atmosphere, and material injected into the atmosphere lowered winter temperatures about 5°C, the kill is not attributed to the acid but to the dry sulfur and fluorine compounds (e.g., Thorarinsson, 1969). These outpourings were not explosive, which is consistent with the great acid release that accompanied them. It is important to note for later discussion that explosive high-silica eruptions (e.g., Santa Maria, Guatemala) are reported not to have large acid yields (Stothers and Rampino, 1983).

Other difficulties have arisen with regard to the impact theory. If an impact did take place, the meteor would not have been a C2 chondrite (Bada and others, 1986). It is now known that the iridium spike in the sedimentary rock at Gubbio is not confined to a section on the order of 1 cm thick, but is distributed over a 4-m-thick sequence astride the K/T boundary. Other chemical elements accompanying the K/T iridium suggest a volcanic origin (e.g., Crocket and others, 1986; Tredoux and others, 1987; Graup, 1987). In addition, gradual extinction histories of selected life forms characterize the end of the Cretaceous (e.g., Zinsmeister and others, 1987); many plants and animals survived into the Tertiary (e.g., Lerbekmo and others, 1987; Buffetaut, 1987; Herman, 1986, Perche-Nielsen, 1987; Hallam, 1987; Kauffman, 1982; McLean, 1985a). These include dinosaurs (e.g. Sloan and others, 1986; Rigby and others, 1987a). If the impact had devastated flora to the extent proposed, then the significant coal measures atop the K/T clay in some areas are rather peculiar. There is also an apparent world-wide hiatus at the K/T boundary (McLean, 1981). Unconformities abound. In some places, as much as the first 7 m.y. of the Tertiary is missing from the stratigraphic record (e.g., Fassett and Obradobich, 1986; Rigby and others, 1986). Other extinct life forms associated with the Cretaceous may have survived well into the Tertiary. It has also been suggested that the impact started great fires, which would

---

Rice, A., 1990, The role of volcanism in K/T extinctions, *in* Lockley, M. G., and Rice, A., eds., Volcanism and fossil biotas: Boulder, Colorado, Geological Society of America, Special Paper 244.

explain the occurrence of soot and charcoal at the K/T boundary (Wolbach and others, 1985). However, such material is common in the mid-Holocene deposits of the Amazon and in beds that considerably pre-date the earliest human presence (Sanford and others, 1985). Apparently, no impacts are involved. This indicates that equally significant fire disturbances may have always been common to the floral environment.

High levels of iridium have not been found in association with other important extinction events, such as the Permian-Triassic (Clark and others, 1986). Although high iridium concentrations have been associated with the Ordovician-Silurian extinctions, there is no iridium spike, and these concentrations are spread throughout a zone representing 20 m.y. It is believed that the high iridium and chromium deposits came from eroded ultramafic rocks rather than a meteor impact (Wilde and others, 1986). Although iridium enrichment is reported for the Mississippian-Pennsylvanian boundary, the Pt-group element-abundance ratios argue for a terrestrial source, not a meteoritic source (Orth and others, 1986). It has been further argued that Ir abundances in the Upper Devonian could just as well have had a terrestrial derivation (Playford and others, 1984). Kauffman and Harrison (Strategies for Mass Extinction Events, seminar at the Department of Geology, University of Colorado, 4 March, 1988) report five layers of iridium at the Cenomanian-Turonian boundary, each separated by approximately 30 cm, but with PGE distributions indicating a deep-mantle source. Koeberl (personal communication, 1988) reports iridium enrichment of volcanic ash found in polar ices. The impact site(s) remains to be found. Originally estimated to be in the ocean where it could be subducted away, the impact must have been ashore or on continental shelves, according to the Si and other elemental distributions in the K/T clay (Bohor and others, 1987). A significant impact site has been located in the continental shelf waters off Newfoundland, but it occurs 10 m.y. too late (Jansa and Pe-Piper, 1987), and no mass extinctions seem to be associated with it.

The "Death Star" that was proposed to have perturbed the Oort cloud to rain meteors and comets down on the inner solar system has yet to be found, and in any case, it would not yield the comet showers as originally hoped (Bailey and others, 1987). MacDougall (1988) suggested that the K/T $^{87}Sr/^{86}Sr$ ratio is the manifestation of intense acid weathering brought on by a flood of atmospheric H_2NO_3 created by bolide impact. Deep Sea Drilling Project and continental data show, however, that the rise in the Sr ratio may have initiated 5 m.y. before the K/T event (Koepnick and others, 1985; Hallam, 1987; Officer and others, 1987), which more nearly reflects the Sr erosion from runoff during the Late Cretaceous regression. Meteoritic acid formation may not have had the impact on life that has been speculated. Rains with pH of 1.7 have been reported in California (Roth and others, 1985), and presumably natural biogenic acid rains of pH 3.6 are common to Alaska (Klinger, 1988a). The death of forests in Alaska is not attributed to the acid rains, but to root kill by encroaching mosses. It is now recognized that bogs are a major source of the acidity in northern climes (e.g., Nriagu and others, 1987). Under conditions of increased moisture, bogs will expand rapidly. For instance, an area in Russia nearly the size of Alaska has become bog in the past 5,000 yr (Klinger, 1988b).

The K/T event is characterized by a sudden and rapid bog formation, which complements the fern spike. Rampino and others (1988) suggested that loss of plankton at the K/T eliminated DMS production vital to cloud formation. The lack of cloud cover was deleterius to the dinosaurs; i.e., they "roasted" to death. Andrae (1988), however, points out that as much, if not more, DMS comes from plants ashore, and the formation of bogs has a positive feedback effect; i.e., they create more cloud cover (Klinger, 1988a, 1988b). Bogs themselves are naturally subject to massive fires and may have been the source of the K/T soot, as well as the extensive coal measures, as indicated by the lack of fossils within them (Klinger, 1988a).

The introduction of the impact theory led to further speculations that the impacts were periodic in nature. A periodicity of some 26 m.y. was inferred from statistical analyses of the paleontological record (e.g., Raup and Sepkoski, 1984), but these inferences have now been shown to be incorrect (Stigler and Wagner, 1987). Kerr (1987) mistakenly reported that the appearance of particles of shocked minerals as large as 100 μm at the K/T boundary constituted an "insoluble problem for advocates of volcanic catastrophe" in explaining the events at the end of the Cretaceous. Kerr also drew from the work of Wilson and Huang (1979), which indicated that grains of such size could be lofted no more than 500 km from the point of explosive volcanism. Kerr stated that only a meteor impact could provide the worldwide dispersal of shocked minerals at the K/T boundary, which has been interpreted by others as proof of impact (e.g., Bohor and others, 1987). Kerr's assessment and those of Wilson and Huang are incorrect. Bailey and others (1984) reported large concentrations of giant particles in the present Arctic atmosphere. These concentrations appear to be uniformly distributed to 5 km in height. Some of these particles are reported to have the appearance of volcanic ash. Concentration does vary with size; i.e., the inverse cube of the particle diameter. From their data, these workers estimate the fall speed of a spherical particle 100 μm in diameter to be ~0.05 m/s, and much lower for plate-like aggregates. These fall speeds are considerably less than the 0.25 m/s for equally large diameters predicted by Wilson and Huang (1979). For that matter, Asian-origin quartz crystals about 300 μm in diameter are known to be transported by wind from Asia to Hawaii (Duce, personal communication, 1987); mineral particles in the 50 μm range can be transported by wind distances in excess of 10,000 km; and particles as large as 100 μm have been transported from the Sahara to Miami, Florida (e.g., Betzer and others, 1986; Duce, 1986; Uematsu and others, 1985; Dauphin, 1983; Carder and others, 1986).

The "microtektites" that attend the K/T boundary are now attributed to fly-ash particles (Cisowski, 1988b), and the microtektite horizons in upper Eocene marine sediments are not associated with mass extinctions (Keller and others, 1983). The proposed impact site for the K/T event, the Manson crater, has now been shown not to have been coincidental with the K/T boundary (Cisowski, 1988a). Because the CO_2 event at the K/T has been an enigma to proponents of the impact theory, it has recently been proposed that the CO_2 event arose from a bolide impact in limestone (O'Keefe and Ahrens, 1988). That paper, from impact proponents, finally acknowledges the greenhouse environment that immediately followed the K/T boundary event. The early Tertiary greenhouse environment will be discussed in detail below, especially as regards its contribution to extinctions, but the O'Keefe and Ahrens (1988) paper again proposed an instantaneous event and continued to ignore paleontological evidence that the extinctions took place over several millions of years. In addition to the injection of CO_2, it can be anticipated that volcanogenic SO_2 would also be discharged into the atmosphere. It would further be anticipated that the SO_2 would contribute to an acid environment, the extent of which is presently confounded by the biotic contribution and the Icelandic experience mentioned above. Further research will sort this out.

In a recent paper proposing that comet showers are the cause for extinctions, Alvarez and others indicated, "it is still too early to predict whether the hypothesis of comet showers will offer a general and global explanation for any or all mass extinctions" and that they do not maintain that this comet showers hypothesis "has been demonstrated to be correct" (Hut and others, 1987).

VOLCANICALLY DERIVED GREENHOUSE-STYLE EXTINCTIONS

McLean (e.g., 1985b) concluded from the paleontological, ^{13}C, and other isotope records that massive CO_2 degassing of the Deccan trap mafic outflows at 65 Ma was a major contributing factor in the extinctions marking the end of the Cretaceous. Oceans of the time were sluggish and warm, from surface to abyssal deeps, hence unable to buffer a 10 to 25 percent increase of atmospheric CO_2, leading to greenhouse conditions (McLean, 1978). A number of workers have cited isotopic evidence confirming a greenhouse effect that might attend such an injection of CO_2 (e.g., McLean, 1985a, b; Perche-Nielson, 1987). There was an apparent atmospheric warming of perhaps as much as 12°C immediately after the end of the Cretaceous, which apparently coincides with the "CO_2 event" characteristic of the K/T boundary (e.g., Margolis and others, 1987). Courtillot and colleagues (1986) confirm from paleomagnetism that the Deccan flood basalts lie directly athwart the K/T boundary; they fix the maximum interval in which the flooding took place to be about 500,000 yr. It has been inferred that the CO_2 output of Iceland through a 20-m.y. period is equal to 60 percent of the present atmospheric CO_2 content, most of which has been absorbed by the oceans (Thorarinsson, 1969). Iceland occupies a surface area of about 100,000 km^2. The Deccan Traps cover about 800,000 km^2. A CO_2 dump of the size proposed by McLean into the biosphere would, over the period of its injection, have caused a continuous drop in the pH of the upper ocean as the ocean took up a portion of the CO_2. The continual decrease in the pH of the upper ocean would selectively terminate one species of microplankton after another, according to their ability to withstand the pH change. For instance, planktonic foraminifera perish when the pH of ocean water drops to 7.8, and coccolithophorids perish at a pH of 7.3 or less (McLean, 1985b). The loss of the coccolithophorids would have disrupted the Williams-Riley CO_2 pump from the atmosphere to seas, disabling an already-feeble ability for the oceans to mitigate the CO_2 injection into the atmosphere. The accompanying loss of biogenic $CaCO_3$ could be one factor by which to explain the K/T boundary hiatus. McLean (personal communication, 1988) indicated the marine extinctions are difficult to attribute to the regression at the end of the Cretaceous, as they were relatively instantaneous when compared to the lengthy time of the sea withdrawal. In addition, the simultaneous response of both benthic and planktonic biota due to the K/T event eliminates the Arctic-water-spillover mechanism originally proposed by Gartner and McGuirk (1979). There is evidence that climate changes due to causes other than meteor impact can engender significant evolutionary changes; the climatic changes brought about by the closing of the Isthmus of Panama are a case in point (Cronin, 1985). Excess CO_2 has been suggested as the cause for dinosaur extinction at the end of the Cretaceous. The literature indicates a number of ways in which excess atmospheric CO_2 could facilitate these extinctions. Oelofsen (1978) noted that atmospheric CO_2/O_2 imbalance due to injection of excess CO_2 from the kimberlite pipe swarms that attended the end of the Cretaceous may have created respiratory difficulties for dinosaur eggs: their small surface-to-volume ratios restrict gaseous diffusion even in the best of circumstances. Recent work by Berner and Landis (1987) suggests that from 75 to 95 Ma, the oxygen concentration of the atmosphere was higher (32 percent O) than it is today (21 percent), which lends to Oelofsen's interpretation. Berner and Landis' results may explain why the reptilian respiratory system is relatively inefficient in comparison to mammals (e.g., Mount, 1979): it evolved in an atmosphere that placed less demand on the cardiovascular system. However, even Cretaceous cardiovascular systems must have had enough elasticity to adapt to an elevation change equivalent to that between the Mexican coast and above the high Sonoran desert, elevations of some 3,500 m. Lizards abound there from bottom to top.

There are abnormal dinosaur eggs from the early Tertiary that show thinning (Erben and others, 1983). If present-day avian species represent a viable analog of early Tertiary physiological response, it is important to note that the eggshells of commercial and other fowl thin considerably during high climatological temperatures (McLean, personal communication, 1986). Internal body temperatures of birds are at least one degree above that of any mammal, recognition of which has led to the suggestion that birds cannot bear live young because their embryos could not be sustained at avian internal body temperatures. The egg must leave the oviduct within 24 hr of ovulation to be viable (e.g., Lewin, 1988). Excess environmental temperatures affect reproduction in animals in other ways. A rise of average ambient temperature of

5°C for sufficient length of time will cause reptiles to lay eggs of a single sex (Ferguson and Joanen, 1982, 1983). The amount of data demonstrating that environmental temperatures a few degrees above optimum for growth can seriously compromise the development of young is astonishingly extensive (e.g., Mount, 1979). Man, however, is a "tropical animal" (Clark and Edholm, 1985) and is probably best adapted to excess environmental temperatures: no hair, the ability to sweat, etc.

The impact theory calls for a nuclear-winter mechanism, whereas the isotope evidence indicates a sudden temperature increase at the K/T boundary. The temperature rose well over that in the last stages of the Cretaceous—10° to 12°C (e.g., McLean, 1985a). These elevated temperatures apparently persisted during a period of about 50,000 yr (e.g., Hsü, 1984; Perche-Nielsen and others, 1982; Boersma and others, 1979; Margolis and others, 1987; Ekdale and Bromley, 1984). In any event, the discovery of dinosaurs well above the Arctic circle (e.g., Brouwers and others, 1987) would indicate the survival rate for a nuclear winter may be much higher than originally envisioned by the impact proponents. In addition, there is much literature indicating that large animals are more cold tolerant than heat tolerant (e.g., Mount, 1979) and adapt much more readily to the cold. Witness the Bengal tigers frolicking in the January snow at the Stockholm zoo. In any case, excess temperatures have an impact on a wide variety of animals. An intra-uterine temperature increase of 1.5°C above normal in modern cows means almost certain fetal mortality. Modern cows become hyperthermic during summer days and suffer if nightly cooling does not return them to optimum body temperature. Milk production decreases by half during hot summer months. European cattle brought into South America during the early days of colonization did not reproduce and died, necessitating the importation of species indigenous to lower latitudes. Conception declines 40 percent in cows if temperatures rise from 20° to 30°C. Warm summers reduce placental blood flow in rabbits by 50 percent. Excess heat (not undernourishment) causes dwarfism as well as skeletal abnormalities in the offspring of many species (e.g., McLean, 1986).

McLean has appealed to the above facts to suggest a cause for one of the greatest extinction episodes in the geologic record: the end of the last Ice Age. It is relevant to note with respect to nuclear winter extinction theories that a large and diverse mammalian population flourished throughout the Ice Age. The Ice Age mammals were huge in comparison to those surviving. This occurred in an environment of restricted food reserves and extreme cold: spruce forests extended well into Texas and Georgia. With the warming that initiated the retreat of the ice sheets came an expansion of living space and food reserves, yet the horse, camel, and mammoth disappeared from North America, as did many other species. During this period of warming, the CO_2 content of the atmosphere rose about 50 percent (e.g., McLean, 1986). There is no recorded meteor impact attending the mass extinctions that accompanied this shift to balmier environments, and with the end of these animals came the end of a nuclear winter environment. Skeletal abnormalities and dwarfism, like that occurring in modern-day animals that are heat stressed in the womb, are common at the end of the Ice Age. It is most interesting to note that mammoths associated with Clovis points (e.g., arrowheads) were dwarfed and had reduced tusks (see, McLean, 1986, for references). Most of the large heat-retaining animals disappeared, and those that did survive, e.g., in Africa, developed appendages to facilitate reduction of core body temperature (elephant ears), took to the water (hippos), and lost body hair. Perhaps the high altitude of the African veldts provided the margin of survival to allow these adaptations.

Given the many physiological similarities between dinosaurs and present-day descendants of Mesozoic life forms, it is logical to pursue the possibility that dinosaurs were subject to similar heat-stress handicaps. The key is the animal's surface-to-volume ratio. McLean argues that early Tertiary warming and the preferential survival of small animals is no coincidence: larger surface-area-to-volume ratio enhances heat removal. Rigby and others (1987a) have shown that dinosaurs surviving into the Tertiary possessed structure that enhanced heat removal. The frills of Ceratopsians are so vascularized (honeycombed with blood vessels) that they do not have the strength to operate as a defensive device. It is obvious from sectioned fossil frills that arterial flow (rounded blood vessels) ran over the outer frill and venous plumbing (flat blood vessels) returned flow to the body core on the underside of the frill. *Triceratops* horns were also vascularized. The jaws of the *Triceratops* jutted out farther than the horns, which would dictate lowering the head to assume an offensive posture. However, the horns would then point directly to the ground instead of ahead, placing the *Triceratops* in the position of having to plow its way to its adversary. Further, any frontal blow would be absorbed off center of the occipital linkage and would likely fracture the neck of the animal. The alignment of the horns with the tops and sides of the frills suggests that, in addition to heat transfer, these appendages served as "cat whiskers," they assisted the animal in feeling its way while foraging, in order not to catch a limb behind the frill and become entrapped. This capability suggests the animal may have been a browser. On sectioning, the tail club of the Ankylosaurus is equally, if not more, vascularized than a Ceratopsian's frills. The club resembles a sponge and lacks the mass and structural strength to be fully effective as a clubbing device. The fact that the tail vertebrae of the Ankylosaurus are fused prohibits flexibility for effective use as a weapon. An analysis of the movements available to the fused tail indicates that, on striking a massive object, it would have a propensity to snap. Swinging appendages can double the rate of heat transfer, and the structure of the Ankylosaurus tail and "club" likely indicates that both swung from side to side as the animal walked. Equally vascularized is the head of the Pachycephalosaurus, whose well-rounded dome hardly argues for its use as a butting device; striking heads would be more likely to glance off one another, without delivering the full impact that could be provided by a flattened forehead. The vast surface areas in the nasal passages of hadrosaurs would readily facilitate heat transfer to regulate core temperatures of these animals. Ankylosaurus, Pachycephalosaurus, and hadrosaurs and Ceratopsians, are the dinosaurs that survived the K/T transition.

Although possessing a much greater heat-transfer advantage over dinosaurs because of their dimunitive size, present-day reptiles can experience excess core temperatures while active and strive to maintain a relatively stable body temperature by moving in and out of the sun, seeking the shade when their core temperatures rise to about 38°C. They remove themselves from the midday heat by burrowing into the sand or retreating underground (e.g., Mount, 1979). The growth of birds is quite responsive to temperatures: excess temperature reduces growth rate and enlarges wattles and combs, ostensibly to facilitate transfer of excess core heat. To maintain a constant body temperature, heat dumped from the surface of the animal to ambient must equal metabolic heat production, i.e.,

$$hA \Delta T + E = qV$$

where A is the surface area of the animal; ΔT is the temperature difference between hide and atmosphere; q is the average metabolic heat production per unit volume; V is the animal volume; E is the evaporative heat loss (respiratory, etc.); and h is the sum of the radiative, conductive, and convective heat-transfer film coefficients. The very effective forced-convective heat transfer of the circulatory system is able to convey core temperatures to the outer appendages. The animal must survive extreme conditions, e.g., 100 percent humidity wherein $E = 0$. Increased ambient temperature diminishes the temperature difference between the animal and surroundings, thereby decreasing the heat transfer. Compensation may occur by increasing skin temperature by epidermal vasodilatation, but only to a point. The metabolic heat production q is often decreased under heat stress in an attempt to avoid excess body core temperature. For a living animal, however, there is always a minimum metabolic heat production. Given a set volume V, the only other compensation left to the animal is to be able to maximize its surface area, which for extremes may be given by $A = qV/h\Delta T$. The dependence of h on temperature in the atmospheric conditions considered here should not be overly large; hence the percent change in effective radiative area needed to compensate for temperature rise will be $A2/A1 = \Delta T1/\Delta T2$. For instance, if $T1$ is halved such that $\Delta T2 = (\Delta T1)/2$, the heat transfer from the animal will be halved; if the animal was initially in equilibrium conditions, its core temperature will rise unless it has twice the effective surface area needed to maintain its initial state of core temperature equilibrium. For example, if the maximum allowable temperature during activity is 36°C and the ambient temperature is 20°C, $\Delta T1$ is 16°C. If the ambient temperature rises 8°C (to 28°C), the animal must have some means to double its effective surface area to maintain the core temperature it had at 20°C, assuming the extreme conditions given above. Vasodilatation into an appendage such as a frill is one way to provide an increase in effective surface area. High temperatures do induce vasodilatation into the horns of modern ungulates (Rigby and others, 1987b), suggesting that the massive horns of the Irish elk may have been an evolutionary attempt to deal with heat stress. During the cooling period near the end of the Cretaceous, increased seasonality favored the survival of dinosaurs with both heat-retentive and heat-radiating capabilities (to survive minimum winter and maximum summer temperature extremes, respectively). Dinosaurs surviving the K/T transition could do both. They were of large volume and had large radiative surfaces whose heat-transfer capabilities could have been modulated by varying the blood flow to them—restricting the blood flow when it was cold and expanding the blood flow when it was warm. It should be noted that young Pachycephalosaurs did not have the pronounced head dome of the adults of the species (Norman, 1985), but their surface-to-volume ratios were considerably greater, making them more efficient heat radiators. Work on the heat-transfer capabilities of the Pelycosaur sail indicates that it was not very effective (Haack, 1986); however, the film coefficient employed in that study appears to be too small by a factor of 50, and the study ignores another on the thermoregulatory nature of insect wings (e.g., Douglas, 1981).

SHOCKED MINERALS: EXPLOSIVE VOLCANISM AS A SOURCE

The question of whether volcanism or impact caused the K/T events revolves around the source of shocked minerals found at the K/T boundary. The discovery of shocked minerals along with iridium at the K/T boundary has assumed crucial importance for the proponents of the impact theory, i.e., if only an impact can generate the pressures to form shocked minerals, the impact theory is relieved of the discovery of the iridium discharges from the Hawaiian volcanoes. It is argued by impact proponents that volcanic explosions will never develop pressures greater than the overlying hydrostatic head. This is because once the hydrostatic head is exceeded, the overlying material will be disrupted and the pressure released (e.g., de Silva and Sharpton, 1988). Only a cursory examination of the explosive literature shows these arguments to be patently false. Pressures to a megabar regularly develop on detonation of TNT in quarry blasting wherein the overburden is only tens of meters thick, providing a restraining pressure of only some tens of bars (e.g., Teller and others, 1968). Certainly disruption occurs, which is precisely what is sought. The megabar pressures still attend. Inquiry indicates that not even confusion between deflagration and detonation is involved. An example of a deflagration is the ignition of unconfined gunpowder, which burns but does not detonate. If the gunpowder is confined, the buildup of pressure while it burns will greatly accelerate the reaction, carrying it to completion so rapidly that overburden and tensile strength pressures are considerably exceeded before the confining material has time to respond dynamically. This leads to explosive failure. However, untamped gunpowder, e.g., gunpowder set in the open air and unconfined, will only flare when it is touched off. Detonation-type explosives are a completely different matter, however. TNT detonated in the open air, i.e., unconfined, will yield open-air pressures in the megabar range. The explosive reaction in TNT or other detonating explosives is not carried forward by burning but by shock. The decompositional reaction front accompanies the shock front and the block of TNT is completely reacted in the time it takes the shock to traverse the TNT. The shock speed in TNT runs about 8 km/s; hence, the TNT is completely decomposed before the evolved gases have even begun to expand (e.g., Cook, 1968). The point is that pressures well in excess of that provided by overburden invariably occur in industrial processes. As indicated below, shock does attend volcanic processes and yields pressures well above the hydrostatic head imposed on the magma chamber.

Williams and McBirney (1979) mention idealized ballistic equations (e.g., assumed no friction, etc.) from which pressures in the tens of kilobar range, attending volcanic explosions, can be deduced. These equations provided very conservative estimates. Had Williams and McBirney acted on their observation concerning these high pressures and corrected for multiphase flow (e.g., Bain and Bonnington, 1970), they would have deduced that pressures in explosive volcanism could run well into the hundreds of kilobar range. The 18 May, 1980, explosion of Mount St. Helens provides similar conclusions, which can be reached by a variety of means.

The seismicity of the 18 May, 1980, eruption recorded at far-field stations indicates a massive explosion occurred at depth perhaps 11 seconds before the failure of the north slope of the mountain and subsequent eruption. Kanamori and others (1984) report complete azimuthal uniformity in far-field P-wave arrivals, first motion up. S-wave amplitude was small. This is the seismic signature of an explosion at depth (e.g., Bolt, 1976; Rodean, 1970). In addition, the north slope of Mount St. Helens failed in three sections. Explosions induce failure by spalling the confining material into slabs. Failure occurs only through the shock accompanying the detonation and not by the expansion of the gases generated by the explosion (e.g., Hino, 1959). Although an explosion does comminute and plastically deform material in a relatively confined region about the detonation center, it is important to emphasize that an explosion does not generate failure from the detonation point outward, but back in from the surface toward the detonation center. Mount St. Helens experienced "retrogressive failure" (Voight and others, 1982). That is, from the outside inward. This is precisely how material fails under explosive attack from within. The shock of the explosion moves out from the detonation point to the surface of the material in which the explosion occurred, and the shock is reflected back toward the detonation center from the interface as a tension wave. As the compression and tension shocks pass each other, proceeding outward and inward, respectively, their stresses will, at some point, exceed the tensile strength of the material, and at that point, the material will fail and a slab of it will separate from the main body. Further slabbing will continue back in toward the detonation point. The thickness of the spalled slabs is about one-half the shock length, which in turn, has length close to the crater depth (e.g., Hino, 1959). Assuming the crater depth at Mount St. Helens to be 1 km, then the north slope should have slabbed into sections approximately 500 m in width if it failed under explosive attack from a detonation at depth. The U.S. Geological Survey reports that the north slope of Mount St. Helens fell in three consecutive sections, each approximately 500 m thick. The failure follows the reflected tension wave back toward the detonation point (see Rice, 1987, 1985, for additional detail).

Moore and Rice (1984) determined that the landslide velocity immediately after the north slope failure at Mount St. Helens had a magnitude of 50 to 80 m/s. Assuming a slope angle of 20° and no friction, it would take ~15 to 25 s to acquire these velocities under the action of gravity alone. The maximum run of the slide would then have to be about 900 m. The landslide is reported to have advanced about twice as far by this time. This establishes that the north slope had an initial "throw velocity" imparted by the shock of the explosion. The throw velocity will be twice the particle velocity accompanying the shock. Assumption of a ~40 m/s throw velocity would be consistent with particle velocities of 20 m/s observed in tuffs under explosive attack (e.g., Larson, 1977). Such a particle velocity implies a shock velocity less than sonic (Teller and others, 1968). In solids, shock may advance at speeds less than that of sound, and shock in solids is often accompanied by an elastic precursor (see Rice, 1985, 1987 for details and references). Immediately before the failure of the north slope, the mountain "rippled and churned" for a number of seconds. This would be indicative of an elastic precursor, which furthers the suspicion of shock-induced failure and suggests a sound speed characteristic of, say alluvium, i.e., 0.5 km/s. This yields an estimate of the shock pressure p at the mountain's surface from

$$p = \rho U v \sim 150 \text{ bar}$$

where v is the particle velocity, U the shock velocity, and the mountain material density is taken as $\rho \sim 1.5 \times 10^3$ kg/m^3 (e.g., Hino, 1959). This pressure will be larger than the anticipated tensile strength of the north slope material. Using $p = 150$ bar, an estimate of the detonation pressure p_D at the explosion site can be obtained from

$$p = p_D (a/r)^2 \sin^2\alpha$$

where the cone angle α of the crater is taken to be ~20°, the reduced radius $a \sim 200$ m, and the depth of the detonation point $r = 4.5$ km. This yields a detonation pressure ~700 kbar, which is more than sufficient to generate shocked minerals. The spike pressure should be about double this, i.e., ~1,400 kbar. The spike pressure is the pressure at the reaction front that accompanies the shock front. The detonation pressure is the pressure at the end of the reaction zone. This estimate is conservative because: (1) some workers are placing the depth of the explosion as far down as 7 km (e.g., Rutherford and others, 1980), and (2) the estimation of north slope throw velocity herein assumed no friction in the slide kinetics. It has not been necessary to employ an explicit equation of state in the above results, although they were obtained by accessing Hugoniots that have been established from considerable experimental work on materials of a geologic nature (e.g., Teller and others, 1968).

It is practice to separate the pressure history of a detonation into several stages for analytical treatment. The most extreme pressure in a detonation is termed the spike pressure (p_S). It accompanies the shock front that proceeds into the undecomposed explosive to ignite it. The spike pressure defines the front of the advancing shock wave and the coinciding front of the reaction zone, within which decomposition takes place. The specific volume of the charge is smallest at the spike pressure, but recovers, i.e., expands, through the reaction zone. The end of the reaction zone defines the Chapman-Jouguet (C-J) plane at which decomposition is complete. The pressure at the C-J plane is defined as the detonation pressure $p_D = (p_S/2)$. Pressure continues to drop, however, and the specific volume continues to expand to the explosion pressure $p_E = (p_D/2)$. The explosion pressure is defined by the return of the volume of the charge and the volume of the inerts to their pre-detonation values. The volume of the inerts here constitutes the co-volume (e.g., Cook, 1968). The explosion pressure is then given by

$$p_E(v - \alpha') = nRT$$

where v is the pre-detonation specific volume, α' the pre-detonation co-volume, n the number of moles, R the gas constant, and T temperature. A CO_2 content of 10 wt% yields an explosion pressure of about 60 kbar. Regardless of the equation of state, the detonation pressure is about twice the explosion pressure (e.g., $P_S/2$; Hino, 1959), i.e., about 120 kbar in this case. The spike pressure should be about 240 kbar. A concomitant water content of 3 wt% would add another 75 kbar or so the detonation pressure, raising the spike pressure to about 400 kbar. The independent results of the slide kinetics given above and the seismic inferences given below suggest the volatile content of the explosive section of the Mount St. Helens magma chamber to be higher than employed here. A CO_2 content of 30 wt% would be more in line with these other results.

Little of the energy of an underground explosion is expended in seismic energy. The coupling coefficient between explosive energy and generated seismic energy runs about 10^{-3}. Most of the energy of an underground explosion goes into inelastic deformation, local brecciation, and phase changes. Kanamori and others (1984) report the magnitude of the Mount St. Helens event to be $M_S = 5.2$. There are numerous relations in the literature equating magnitude of earthquake with energy (e.g., see Kasahara, 1981). Bath's relationship, i.e.,

$$\text{Log } E_S = 1.44 M_S + 12.24$$

provides an E_S of 6.0×10^{14} J for Mount St. Helens. The blasting literature (e.g., Cook, 1968) indicates that the magma-chamber wall displacement will be of the order of meters in such case. This yields pressures in the megabar range. The discovery of shocked minerals in volcanic discharge (e.g., Carter and others, 1986, 1989), in particular the Mount St. Helens ash that fell near Pasco, Washington, support the above assessments.

Although these points should leave no doubt that extreme pressures attended the Mount St. Helens failure, it is comforting to some investigators if a mechanism can be proposed that would yield such pressures. For reasons outlined in Appendix I, melt-coolant interactions are eliminated from consideration. In addition, Appendix I shows that geologic phenomena attributed to melt-coolant (or fuel-coolant) interactions may have other interpretations. An interesting process that is known to yield high pressures is suggested here. It is common industrial experience for solidifying melts to develop very high volatile pressures within them, even to the point of bursting through several centimeters of solid steel. This phenomenin is depicted in Figure 1, which also shows the postulated carry-over to the magmatic environment. When an ingot of steel is poured and set to cool, a quenched layer is immediately frozen to the sides, bottom and top of the container. The quench margin impedes heat transfer and slows further cooling and freezing of the interior. Note the solid cap at the top of the ingot container, which does not founder on forming even though it is heavier than the underlying melt. Quenching the surface melt of the ingot greatly increases the temperature-dependent viscosity of the melt at the top and also generates innumerable small crystals there, which so thickens and immobilizes the surface melt that it solidifies and becomes competent before it can sink. Industrial melts almost invariably contain volatiles. However, volatile solubility is many orders of magnitude smaller in the solid phase than in the liquid phase. As the ingot freezes inward, the volatile content is pushed into the remaining melt, supercharging the remaining melt with gas. The final melt becomes supercooled before freezing (as do all melts). Hence the last melt ends up supersaturated in gas and liquid phases. Further cooling to the nucleation temperature causes loss of this metastable state with the formation of a crystal nucleation site that initiates the freezing. Volatiles are ejected from this site into a high-temperature environment too quickly to be diffused into a new phase (diffusion times in silicic melts are on the order of geologic time scales). This generates a local pressure spike that provides activation energy to generate more nucleation sites, which in turn, generate more gas; this autocatalytic chain reaction is driven by the pressure associated with the dumping of the gases from the melt. This pressure release sweeps through the melt as a shock front, spawning solidification and further gas release. The volatiles in the melt are dumped within the time it takes for the shock to traverse the remaining melt. This occurs in the same time frame as observed in industrial melts: just about instantaneously (see Rice, 1985, for references to the industrial literature).

When applying the industrial experience to the geologic environment, it is important to note that fluids possessing both a temperature gradient and a concentration gradient break into layers of different density, the density of each layer increasing with depth. The convecting system in the fluid is, therefore, one of

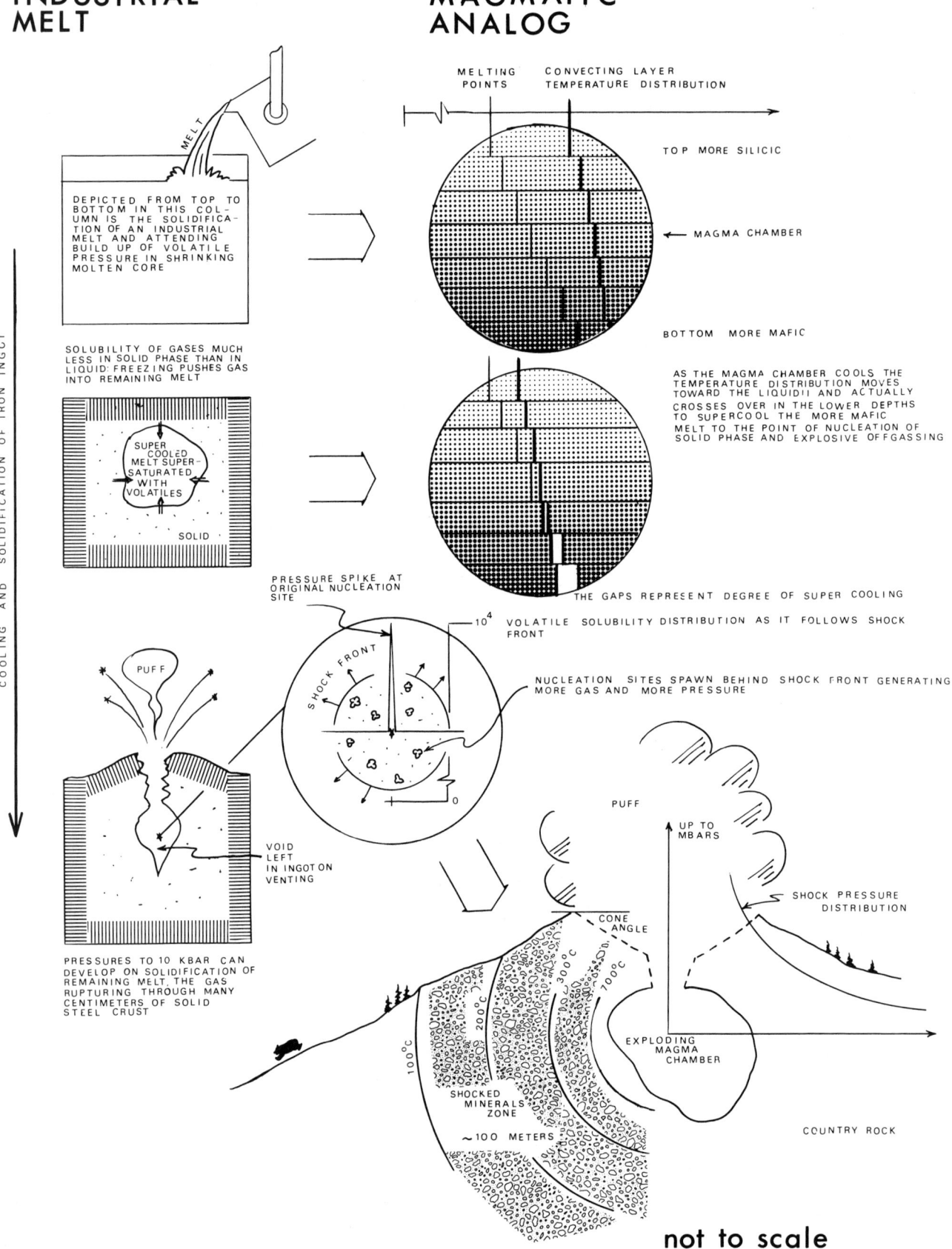

many layers; each layer is so well mixed that it is of uniform composition and temperature throughout. A fluid of uniform composition would compose a single layer bounded only by the top and bottom of the fluid. In the case of varying composition, there are gradients only at the boundary layers separating each layer of convection. This gives the vertical variation in temperature and composition a staircase-like distribution, both temperature and mafic content increasing with depth. A representative temperature profile is shown in Figure 1, as well as a representative melting point distribution in the magma chamber. The more silicic material at the top of the chamber has a lower melting point than the more mafic material near the bottom; the variation in composition is shown by the grading in the idealized representation of the magma chamber. As the magma chamber cools, the staircase-like temperature distribution shifts to lower tempera-

◄─────────────────────────────────

Figure 1. The quench supersaturation mechanism, modeled after industrial experience as ingots are poured in a foundry is shown (left). A quench margin immediately forms a rim at the walls and surface of the ingot. Note that a solid skin forms over the top of the melt without foundering, even though the solid phase is heavier than the liquid phase. The rapid cooling of the surface melt quickly increases its viscosity. The attendant generation of many small crystals at the top of the fluid further thickens the surface melt so that by the time it can move to sink it is already solidified in place and at the walls. The quench margin impedes heat transfer, slowing the solidification within. The volatile solubility of the solid phase is many orders of magnitude smaller than that of the liquid phase; hence, as the solidification front advances into the interior of the ingot, the volatiles are pushed into the remaining melt, supersaturating the melt with gas. The melt eventually becomes uniformly undercooled and ready to freeze. A nucleation site finally forms within, precipitating volatiles into a high-temperature environment as it does so. This exsolution of volatiles generates a local pressure spike that provides the activation energy to generate more nucleation sites, which exsolve more gas. This runaway autocatalytic reaction sweeps the remaining melt as a shock wave and spawns solidification and gas. The entire volatile content is dumped from the melt in the time it takes for the shock to sweep the remaining melt. Pressures as great as 10 kbar are known to suddenly develop in ingots, bowing up or bursting through several cm of solidified steel.

The effects of the processes occurring within the ingot on the magma chamber are shown (right). A zoned magma chamber is depicted: more silicic material is at the top, more mafic material at the bottom. Convecting fluids with both temperature and concentration gradients in the vertical will break up into layers of convection: each layer well mixed and uniform in composition and temperature throughout, with gradients occurring only at the boundaries separating each layer. This leads to staircase-like distributions; one is shown for temperature increasing with depth. The temperature coordinate runs to the right. Also shown is the distribution of melting temperatures, which also increase with depth; the more mafic material has the highest melting point. As the magma chamber cools, the temperature distribution shifts toward the melting-point temperatures until near the bottom of the magma chamber; the layers actually become undercooled. If the sides of the chamber are glassy, there are no nucleation sites and a hazardous situation arises: the melt becomes so undercooled that explosive crystallization causes release of gas in catastrophic proportions. Shocked minerals will be formed up to several hundred meters away from the magma chamber walls, but those formed within the high temperatures of the magma or the walls are expected to anneal out.

tures. Note that the lower portions of the chamber become cooled below their melting point, setting up a situation similar to that in the ingot depicted on the right. In the magma chamber case, however, it is a layer in the lower portion of the magma chamber that explodes. Off-gassing flash cools other layers in the vicinity, which can result in explosions by these layers also. There were apparently two explosions at Mount St. Helens, about two minutes apart. In industry, mechanical disturbance such as vigorous stirring of supercooled melts does not lead to precipitation of the solid phase, whereas crystal seeding will. For instance, a supercooled metallic melt can be vigorously stirred with a ceramic rod without initiating freezing. However, pitching a nail into the melt will instantaneously lead to massive precipitation of solid phase and explosive exsolution of volatiles (see Rice, 1985, for further detail and references to the industrial literature). Rayleigh number considerations suggest that the convecting layers in the magma chamber are on the order of 10 m thick, which is the same estimate provided by the blasting literature for the thickness of the exploding layer. The above mechanism, which is suggested to apply in magma chambers, is an outgrowth of earlier considerations (e.g., Rice and Eichelberger, 1976; Rice, 1982).

Once open to atmospheric conditions, the magma chamber may continue to off-gas in a "boiling" mode rather than an explosive one, and continue to boil for some time after the explosion, which exposed magma to atmospheric conditions.

It is unlikely that shock features would be retained in phenocrysts or other material in the magma chamber due to the elevated temperatures of the magma, which would anneal out such features. The same applies to the magma chamber walls. However, farther away, where it is cooler, such features could be retained. Rice (1987) estimated the shock features to be derived from surrounding country rock, as far as several hundreds of meters away from the magma chamber. Jointing cracks or angularities in country rock can serve as stress concentrators, which would extend this region considerably farther. If the country rock is not of magmatic composition, shocked minerals from it would not reflect magmatic origins.

The mass transfer associated with the off-gassing of solidifying melts can be quantified from relations developed in industry. For example, the transport coefficient K is given by

$$K_w \propto J^{1/2} V_B D^{1/2}$$

where V_B *is the bubble volume, D* is the diffusion coefficient, and the rate of bubble formation is given by

$$J = z[\exp(-\Delta H/kT) \sqrt{6\gamma/(3-b)\pi m}] [\exp\{-16\pi\gamma^3/3kT(p_C - p_a)^2\}]$$

where $b = (p_C - p_a)/p_C$, z is the number of exsolving species per cm^3, ΔH is the heat of formation of one molecule of vapor from the melt, m is the mass of the vapor molecule, k is Boltzmann's constant, P_c is the pressure in a bubble of critical size such that the partial pressure of the gas in equilibrium with the fluid exceeds the sum of the ambient pressure P_a and the pressure due to the

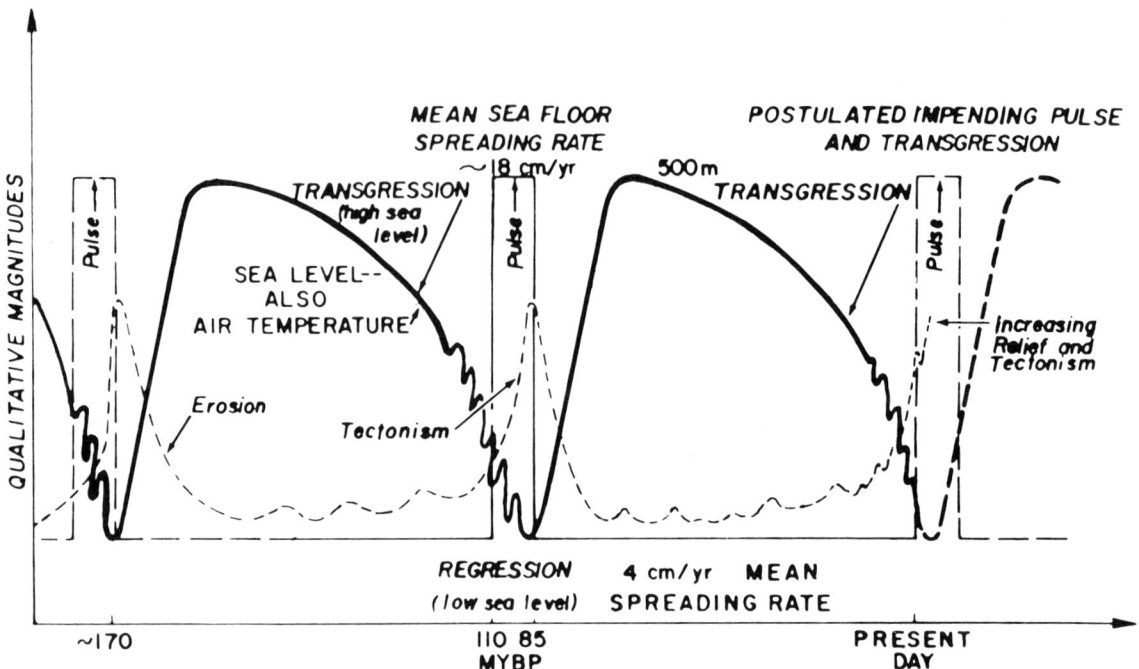

Figure 2. High Rayleigh number mantle convection (whole-mantle convection) can experience periodic surges. If the mantle is composed of a number of convection cells, each with its own periodicity, their superimposition may indicate episodic tectonism, volcanism, and sea-floor-spreading rate pulses.

bubble surface tension, $2\gamma/r$. A 10°C drop in temperature can lead to a 10^{32} increase in the rate of bubble formation (e.g., Katz and Sliepovich, 1971). Gases are known to disassociate in industrial melts, e.g., $CO_2 \rightarrow CO + O$. For magmatic melts, and assuming a formation heat of $O + CO \rightarrow CO_2$ from the melt of about 70 kcal/mole, the overpressure from ambient is about 500 kbar. Shock in industrial melts can also disassociate CO into C and O (e.g., Nellis and others, 1981), which may provide an explanation for the association of soot with the K/T boundary.

Quench supersaturation explosions, i.e., the mechanism described above, is akin to "second boiling" as described by Morey (1922) wherein high pressures may be generated by exsolving gases within a solidifying melt. Although their work has little relevance to K-T events, Loper and McCartney (1988) have brought attention to other work wherein pressure in a magmatic melt was related to the degree of undercooling in an exponential fashion. Pressures to 35 kbar were measured in supercooled igneous melts, the limitations of the experimental equipment preventing the observation of higher pressures. Quench supersaturation explosions, as described above, may have analogs in other types of explosive phase changes (see Appendix II).

The activation energy that is necessary to initiate freezing (i.e., the sum of the surface energy of the nucleation site and the difference in the free energy in the volume of the nucleation site between the liquid and solid state; Knight, 1967) may be given by

$$G = 16\pi\gamma^3 T_M^3/3 \, (L_M^2 \, \Delta T^2)$$

where T_M is the melting point, L_M *is the latent heat,* and ΔT is the degree of undercooling (Davies, 1973) from which it may be inferred that an overpressure on the order of kbar will be sufficient to drive explosive crystallization of mafic magmatic melt. In this context, it is useful to recall the usual form of free energy, $dh-SdT$, where the enthalpy dH is given by $dH = dQ + VdP$. It is to be noted that the accompanying reaction rate equation (e.g., Rice, 1988) yields an exponentially increasing pressure curve as given in Loper and McCartney (1988).

The importance of CO_2 in generating volcanically derived pressures sufficient to form shocked minerals and the discovery of CO_2 inclusions along shock lamellae in minerals of the Vredefort and Subdury (Medenbach and others, 1987; Fricke and Schreyer, 1987) complement suggestions that CO_2 is crucial to a number of important geologic processes (e.g., Nicolaysen, 1985). Augmenting this view is the proposal that the Deccan Trap represented mantle CO_2 outgassing that led to a greenhouse effect, which was deleterious to large-body life forms (e.g., McLean, 1985a, b). The Deccan Trap is now known to lie athwart the K/T boundary, strengthening the suggestion that these massive flows may be a source of the K/T iridium (e.g., Vandamme and others, 1986). The extensive kimberlite pipe fields emplaced at the end of the Cretaceous could also provide an atmospheric CO_2 dump sufficient to overload oceans, which at that time had little buffering capacity. It has been argued that, although the mafic flows of the Deccan Trap may be a source for the iridium, they could not be a source for the shocked minerals. There is, however, a commonality between mafic flows and silicic flows, which could be the source of explosive volcanism, and this did exist with the emplacement of the Deccan Trap (e.g., Lightfoot and others, 1987; Sethna and Battiwala, 1977). Sukheswala and Sethna (1962) note the appearance of patchy extinctions in phenocrysts in silicic outpourings associated with the Deccan Trap. Patchy extinctions are characteristic of mosaicism, a diagnostic feature of extreme shock (Carter and others, 1986, 1989). Although mafic outpourings themselves are not to be an expected source of explosive volcanism, the converse is not necessarily true. Iridium can be expected from explosive volcanism. Note in Figure 1 that the zoned magma chamber is mafic at the bottom and, hence, could also be a source of iridium.

MANTLE DYNAMICS AND EPISODIC EXTINCTIONS

Although recent attention has been directed mainly toward the Deccan Trap as the source of K/T volcanism, many workers note that rearrangements in tectonics are global and not localized, correlating with world-wide variations in sea-floor-spreading rates, which in turn correlate with taxonomic diversity (e.g., Rich and others, 1986). Spreading rates in the Maastrichtian were several times that of today, in both the North Pacific and North Atlantic, and spreading ridges were likely a source of K/T CO_2. Berner and others (1983) have shown that global CO_2 levels are strongly connected to sea-floor-spreading rates, and peaks in spreading rates correspond to peaks in global temperature. Hays and Pitman (1973) and Pitman (1978) have related the Late Cretaceous and Eocene transgressions with a jump in spreading rates, attributing the sea-level rise to increased volume of the mid-ocean ridges. This excess volume was ostensibly the result of a massive influx of mantle material. The Eocene CO_2 greenhouse (a warming of about 5°C) has been connected with a five-fold increase of sea-floor hydrothermal activity (Owen and Rea, 1985). Time variation in global plate motion correlates with volcanic activity. Volcanic activity at the close of the Cretaceous was widespread, e.g., Antarctica (Borella, 1984), South Africa (Oelofsen, 1978), the Deccan Traps in India. Loper and McCartney (1988) have noted that whole-mantle convection would access the core-mantle boundary to provide source material for the iridium; these workers have proposed that the Deccan Trap was a manifestation of a mantle plume. The worldwide occurrence of K/T volcanism invokes a more global mechanism. For example, Kennett (1988) reports volcanic ash 50 cm thick lying exactly at the nannoplankton K/T event—the ash apparently from Antarctic volcanism. It is probable that there is communication between upper and lower mantle (Fischer and others, 1986), which argues for whole-mantle convection. Rice (1972, 1982) and Rice and Fairbridge (1975) have argued that whole-mantle convection provides Rayleigh numbers large enough to secure periodic flow in mantle convection cells. The overprint of the periodicities of a number of cells would appear episodic in the geologic record. A surge in mantle convection would manifest itself in increased sea-floor spreading and transgression, as shown in Figure 2, with follow-on regression and orogenic activity as detailed in Rice and Fairbridge (1975).

APPENDIX I

Kanamori (personal communication, 1987) has adhered to the original interpretation offered by the U.S. Geological Survey concerning the mechanism of the Mount St. Helens eruption, i.e., the release of overburden due to the failure of the north slope ostensibly led to the flashing of the volatiles in the magma chamber (e.g., U.S.G.S. Professional Paper 1250). The lateral exhaust through the side of the volcano was taken by the U.S. Geological Survey to be a horizontal directed blast. In analyzing the near-field seismic data, Kanamori and Gvien (1982) indicated that the arrival of only Love and Rayleigh waves was consistent with this interpretation. It should be recognized that landslides by themselves are effective Love wave generators (see Rice, 1985, for references), and that inversions and source-characteristic determinations are not unique. Further, explosive sources are accompanied by a local shadow zone, with respect to body waves, and the lack of significant P arrivals at near-field stations is consistent with an initiating explosion such as Mount St. Helens. Regardless of views on the cause of the explosion, there is concurrence that the M = 5.2 earthquake preceeded the failure of the north slope by approximately 11 s, which is consistent with a detonation point 4.5 km deep and shock propagation velocities of 0.5 km/s (used here). Kanamori's analyses show the earthquake to preceed the north slope failure (see Kanamori and others, 1984, text and Fig. 16). Kanamori (personal communication, 1988) believes the landslide occurred before the earthquakes, but I am unable to establish a means of arriving at that conclusion. The U.S.G.S. considers the earthquake responsible for the loss of the north slope (e.g., Voight, 1982).

The U.S.G.S. reports Mount St. Helens to have been quiet in the early morning of 18 May 1980, until the "magnitude 5+ earthquake at 0832:11.4 PDT . . . started minor rock and ice falls from the south crater wall . . . Following an interval of a few seconds, a major fracture propagated rapidly along the apex of the bulge north of the summit crater. North of this fracture, the rock rippled and churned, apparently in place, for an interval of several additional seconds. The north face then slid down in a gigantic rockslide." Before collapsing, the surfaces of quarry benches similarly ripple and churn when shock from an explosion below impinges on the air-rock interface (e.g., high-speed filming of quarry blasting, Martin Marietta, courtesy of S. Winzer). The onset of the rockslide began at 0832:21 PDT, with a possible several second error associated with the estimated delay time (e.g., U.S.G.S. Professional paper 1250, p. 81). This clearly places the earthquake before the landslide. Rice (1985) has shown that the north slope landslide of Mount St. Helens did not have characteristics of a gravitational failure. The run and coverage were far too great, and precursors like accelerating downward creep were absent. The significant pre-avalanche deformation that did occur, i.e., that of the bulge, was upward and outward, and was of a de-accelerating nature (Voight and others, 1982). As the internal angle of friction of north slope material greatly exceeded the slope angle, it was impossible for the north slope to have fallen by itself. It required disruption (see Rice, 1985, for detail and references to the earth mechanics (literature). Further, the median slope inclination for earthquake-generated slides is 50°, the minimum slope inclination being 25° (Keefer, 1984). The inclination of the Mount St. Helens north slope was 18 to 23°. Beneath all the rubble and overburden of what had been the north slope, and 2 min after the initial earthquake, there was another magnitude 5, explosive-source earthquake and subsequent explosive venting. This contradicts the release of overburden as an explanation for volcanic explosions. The necessity of uncapping magma chambers to engender explosions leaves unexplained events in which no slide was involved, such as Lamington, New Guinea. The top of Lamington was blown off uniformly, such that the resulting crater is completely symmetrical (P. Lippman, personal communication, 1985). Slides are not involved in the explosions at Sakurajima (Ishihara, 1985), wherein explosion-source earthquakes preceed explosive eruption of the crater floor by several seconds. These observations greatly strengthen my belief that explosion in a magma chamber was the initiating event on 18 May 1980, at Mount St. Helens. Further, that explosion has been placed as deep as 7 km (Rutherford and others, 1980), which makes it even more difficult to believe that the removal of north slope overburden initiated it.

The difficulties in generating melt-coolant explosions or explosions by uncapping magma chambers are briefly reviewed here (see Rice, 1985, for detail and references) but have pertinence in other volcanic contexts. The preponderance of industrial literature indicates that explosive melt-coolant interactions take place in less than a millisecond. If the explosion is due to heat transfer from the melt to the coolant, then intimate mixing must occur within this time frame, which requires, among other things, that the Weber number be large enough to break the melt into droplets smaller than 1 mm in diameter. This would give a surface-to-volume ratio high enough to assure fast heat transfer to meet observed detonation time constraints. These observations negate speculation that lava flows into the ocean will yield steam explosions of hydrogen bomb proportions, as was feared might take place at Heimaey (Kanamori, personal communication, 1986). Constraints imposed by industry will not be met by naturally occurring volumes of magma. Pillow basalts are not millimeter scale, and their occurrence argues against explosive melt-coolant interaction between magma and sea water. Otherwise, they would have been blown to bits. There are further difficulties. Water above 80°C will not explode, no matter how high the temperature of the melt. If the temperature of the melt is extremely high, there will be no explosion, regardless of the water temperature (e.g., Dulleforce and others, 1976). In addition, dry melts, i.e., melts sparged of volatiles, never explode.

The hosing down of white-hot slag to granulate it is a case in point. This has led to the suggestion that the cause of these explosions is quench supersaturation (e.g., Rice, 1985), wherein the explosion is caused by explosive off-gassing during solidification. The quench supersaturation mechanism will be discussed here in further detail, but it will provide explanation for the observation that explosions will not occur for coolants too hot or melts too hot. In both cases it is likely that film boiling occurs, which greatly inhibits the heat transfer from melt to coolant (by factors of 1,000 or more). This allows the melt to cool slowly in the coolant, rather than be quenched, which allows volatiles to bleed off quietly. Quench supersaturation provides an explanation for the vast amounts of lava that enter the sea from Hawaii without generating an explosive response. Scuba divers have filmed red-hot magma advancing across the sea floor underwater. Examples may be seen in the documentary series about Hawaii *Nature*.) No gas was generated by red-hot lava under water. Only until the lava had cooled to a solidified hard crust did gas begin to evolve, and this was by bursting through the crust, i.e., the gas was generated inside the magma as it solidified.

The above constraints will apply even more rigidly to ground water. The effective "viscosity" of sedimentary materials is considerably greater than that of water and, thereby, impedes the rapid intimate mixing of magma and pore water required. This is relevant to a proposed explosion source for a peculiar earthquake off Tiro Shima, which was believed to have been caused by the flashing of water in sea-floor sediments on contact with magma (e.g., Kanamori and others, 1986). The magma will pond between basement and sea-floor sediments because its density probably will be somewhat larger than that of the overlying sediments. A necessary but not sufficient condition for the magma to penetrate the overlying sediments is the rough engineering rule of thumb given by (see Rice, 1981, for references)

$$V > \sqrt{2gh \, \Delta\rho/\rho e}$$

where V is the upward velocity of the magma and h the depth of the overlying material. As this estimate excludes yield strengths, virtual mass, viscosity, etc., it is extremely conservative and, for a sedimentary overburden 3 km thick indicates a penetration velocity considerably more than 2.5×10^2 m/s to avoid ponding. It is unlikely that sediment or melt will break into millimeter-size particles within a millisecond, as required by the literature for an explosive interaction. Argillaceous sediments this thick, as found in the Gulf Coast, would have interlayer and pore water squeezed out of them at the basement boundary and may be claystone. It would take perhaps an hour at 1,100°C to "cook out" structural water at atmospheric pressure, and the temperature would need to increase with depth/pressure (R. Reynolds, Department of Geology, Dartmouth College, personal communication, 1987). The influx of magma at the sediment-basement boundary will hydrofracture the interface, allowing the magma to flood along the basement sediment boundary. This is because the resisting horizontal stress $\sigma_H = K_o \, \sigma_v$ (where σ_v is the hydrostatic head) will likely have K_o very close to zero for sedimentary materials. Even if $K_o = 1$, the vertical hydrofracturing would be small in comparison to horizontal hydrofracturing (N. Y. Chang, Department of Civil Engineering, University of Colorado, Denver, personal communication, 1987). Therefore, the predominate source motion would be the lifting of overburden, which is consistent with the predominance of the SV signal in comparison to P and SH motion reported for this earthquake. These analyses are from LP data alone, which suggests that there was little SP component in the signal, which in turn, would be strong evidence that an explosion was not involved. If little SP component attended the signal, then a displacement with slow rise time is suggested. Kasahara (1981) shows sources with slow rise times that are initially rapid but then taper off.

It is unusual for an earthquake of magnitude 5.5 to generate tsunamis of the size reported for Tiro Shima, unless it is a source with a rise time of some length, e.g., 100 s (e.g., Kanamori, 1972). It is likely that an overpressure above ambient by only several bars will cause separation of sediment from basement. If a dike is feeding the hydrofracture between basement and overlying sediment, an estimate of the time necessary to inject magma to a certain thickness may be obtained from

$$\Delta p = \rho \, (V^2/2) \, (L/D) f$$

where V is the average flow velocity, ρ is the melt density, L the crack length, and D the hydraulic diameter (taken here to be half the thickness of the injected magma). The flow should stop when the injected sill is thick enough to overcome the overpressure above local hydrostatic conditions, i.e., 5 to 10 m thick. A large crack implies a small roughness ratio, and iteration indicates a Reynolds number in the turbulent regime, suggesting a friction factor $f = 0.01$. This suggests that a horizontal crack extending 1 km to either side of the dike could be filled to about 10 m thick in somewhat more than 1 min. This type of filling may also provide an explanation for the peculiar directivity of the radiation pattern for this earthquake. Using $E = A \rho dz$ to estimate the energy release, the above estimate of dz, i.e., 5 to 10 m and an overpressure above ambient by several bars, yields an energy expenditure of 10^{20} dyne cm, if the affected area is about 6 to 7 km². This assumes that the sill extended horizontally 1 km to either side of the crack, i.e., was 2 km across. Assuming the flooded area to be about 3 km long provides the affected area used above. Using the relation

$$\log E_S = 1.44 \, M + 12.24$$

for this magnitude 5.5 earthquake yields an energy release of 1.5×10^{20} ergs, similar to the first estimation above.

Kanamori and others (1986) indicate that the source of the Tiro Shima earthquake was a compensated linear doublet without a couple, hence the moment Mo of this earthquake should be close to the above estimates of the energy release. Kanamori and others (1986) report, however, a moment four orders of magnitude greater. The details of their estimation of moment are not spelled out, but Kanamori and Anderson (1975) provide a graphic relation of M_S versus M_o. Kanamori and others (1986) moment versus magnitude places one at the extreme left-hand edge of the graphic relation provided by Kanamori and Anderson (1975); hence, the curve cannot be followed out to lower moments. Further, the value of shear modulus assumed for this graphic relation is 3×10^{11} dyne/cm^2, whereas that of sea-floor clays is about eight orders of magnitude less at a consolidation stress of ~370 kPa and at 5 percent shear strength. The deepest sediments will lie closer to the value used in the graph, but if the following relation (e.g., Pilant, 1979) is employed,

$$\log M_o = 1.5 M_S + 11.8 - \log(\eta \bar{\sigma}/\mu)$$

we obtain a moment without couple closer to that of the above energy estimates, i.e., $\sim 10^{20}$ dyne cm^2. This assumes a seismic efficiency $\eta \sim 1$ (because lifting, not sliding, is the form of displacement, little friction is assumed to be encountered in opening the crack). It also assumes $\bar{\sigma} \sim 1$ bar and $\mu \sim 10^6$ dyne/cm^2. Hanks and Thatcher (1972) indicate a moment of this value to have a source dimension of 1 km or so, which supports the affected area assumed above, as do the empirical formulations of Utsu and Seki (1955) and others. Little information is given regarding the characteristics of the tsunami associated with the Tiro Shima earthquake. Tide gauges do not often reflect the full height of a seismic sea wave. Apparently, the "run up" on the beach did not yield a bore of great height. A conservative estimate would make the maximum stable height of the wave 1.5 m, suggesting a wave height of 1 m just before landfall. Assuming constant depth d, a rough guess as to the initial wave height might be $H/K_S K_f = H_o$; where $H_o = 1$ m, K_S is the shoaling factor, and $K_f = fH'_o K_S \Delta x/d$; where f is the friction factor, H'_o = the equivalent deep-water wavelength, and Δx (which is normally the fetch) is the projected source dimension on the sea surface. Values from Brietschneider (1969) suggest an initial wave height of 2 m. If the directivity of the sediment displacement of 5 to 10 m is preserved, such an initial wave height is plausible. This discussion makes the point that explosive volcanism is unique and that other mechanisms can be explored if explosive-volcanism characteristics are not within the seismic response.

APPENDIX II

Superheat limit explosions (SLE) are phase-change processes. However at near-surface magma chamber pressures; there is only a degree or so separating the phase separation line and the spinodal (Rice, 1985) hence little superheat. Deeper magma chambers place pressures above the critical point and out of SLE-phenomena range. There are explosive solid-solid phase-change phenomena that propagate up to several hundreds of m/s, but these represent the "explosive" transition from glassy to crystalline metal (e.g., Messier and others, 1975; Shklovskii and Druinskii, 1986). Crystallization velocities on the order of 100 m/s are known to occur in supercooled liquid metals (e.g., Schleip and others, 1987). The process considered here, quench supersaturation, seems closer to "burst martensitic" transformations, which occur, for instance, when attempting to cool from austenitic steel (γ-Fe) into the cementite-ferrite field (α-Fe). The α-Fe is incapable of dissolving all the carbon present in the γ-Fe; therefore, completion of the phase transformation requires the diffusion of C out of γ-Fe and its precipitation elsewhere as iron carbide. If the γ-Fe is quenched, there is no time for this diffusion process, and the carbon gets popped out of the γ-Fe, stranded or trapped in a nonequilibrium solid-state matrix that is not one phase or the other. This matrix is termed "martensite" (e.g., Avner, 1979). Its formation can occur so rapidly that shock is produced in sufficient strength to shatter Dewar flasks containing liquid nitrogen into which the samples were immersed (e.g., Reed-Hill, 1978). The γ-Fe to martensite transition front, which is also the shock front, propagates at several km/s (e.g., Entwisle, 1971). It is important to note that phase diagrams do not have immediate meaning for the above, highly nonequilibrium process; however, they do indicate where the material may eventually end up, given a sufficient length of diffusion time. The martensitic state may have lifetimes of centuries or millenia, if there is no annealing. Strong impact (as opposed to stirring) can induce sudden freezing (e.g., Lovett and others, 1982).

REFERENCES CITED

Alvarez, L. W., Alvarez, W., Asaro, F., and Michel, H. V., 1980, Extraterrestrial cause for the Cretaceous–Tertiary extinction: Science, v. 208, p. 1095–1108.

Andrae, M. O., 1988, Geophysical interactions in the global sulfur cycle; Chapman Conference on the Gaia Hypothesis, San Diego, March 7–11: American Geophysical Union, p. 18–19.

Avner, S. H., 1979, Introduction to physical metallurgy: New York, McGraw-Hill, 696 p.

Bada, J., Zhao, M., Roach, M., and Zare, R., 1986, Amino acids in K-T boundary sediments: Geological Society of America Abstracts with Programs, v. 18, p. 531.

Bailey, I. H., Radke, L. F., Lyons, J. H., and Hobbs, P. V., 1984, Airborne observations of Arctic aerosols; 2, Giant particles: Geophysical Research Letters, v. 11, p. 397–400.

Bailey, M. E., Wilkinson, D. A., and Wolfendale, A. W., 1987, Can episodic comet showers explain the 30-Myr cyclicity in the terrestrial record?: Monthly Notices of the Royal Astronomical Society, v. 227, p. 863–885.

Bain, A. G., and Bonnington, S. T., 1970, The hydraulic transport of solids by pipeline: New York, Pergamon Press, 251 p.

Berner, R., and Landis, P., 1987, The major gas composition of ancient air; Analysis of gas bubble inclusions in fossil amber: New Haven, Connecticut, Yale University, Department of Geology and Geophysics preprint.

Berner, R. A., Lasaga, A. C., and Garrels, R. M., 1983, The carbonate-silicate geochemical cycle and its effect on atmospheric carbon dioxide over the past 100 million years: American Journal of Science, v. 283, p. 641–683.

Betzer, P. R., Bernstein, R. E., Carder, K. L., Breland, J. B., Duce, R. A., Uematsu, M., and Feely, R. A., 1986, Particle fluxes in the North Pacific Ocean; Responses to major atmospheric dust storms: EOS Transactions of the American Geophysical Union, v. 67, p. 899.

Boersma, A., Shackleton, N., Hall, M., and Given, Q., 1979, Carbon and oxygen isotope records at DSDP site 384 (North Atlantic) and some Paleocene paleotemperatures and carbon isotope variations in the Atlantic Ocean *in* Initial reports of the Deep Sea Drilling Project: Washington, D.C., U.S. Government Printing Office, v. 43, p. 695–717.

Bohor, B. F., Modreski, P. J., and Foord, E. E., 1987, Shocked quartz in the Cretaceous–Tertiary boundary clays; Evidence for a global distribution: Science, v. 236, p. 705–709.

Bolt, B., 1976, Nuclear explosions and earthquakes: San Francisco, California, W. H. Freeman Co., 309 p.

Borella, P. E., 1984, Initial report of the Deep Sea Drilling Project: Washington, D.C., U.S. Government Printing Office, v. 74, p. 645–652.

Breitschneider, C. L., 1969, Sea motion, *in* Meyers, J., ed., Handbook of ocean and underwater engineering: New York, McGraw-Hill, 890 p.

Brouwers, E. M., Clemens, W. A., Spicer, R. A., Ager, T. A., Carter, L. D., and Sliter, W. V., 1987, Dinosaurs on the North Slope, Alaska; High latitude, latest Cretaceous environments: Science, v. 237, p. 1608–1611.

Buffetaut, E., 1987, Vertebrates and Cretaceous–Tertiary boundary events; International Workshop on Cryptoexplosions and Catastrophes in the Geologic Record, 6-10 July, Parys, South Africa: University of the Witwatersrand.

Carder, K. L., Steward, R. G., Betzer, P. R., Johnson, D. L., and Prospero, J. M., 1986, Dynamics and composition of particles from an aeolian input event to the Sargasso Sea: Journal of Geophysical Research, v. 91, p. 1055–1066.

Carter, N. L., Officer, C. B., Chesner, C. A., and Rose, W. I., 1986, Dynamic deformation of volcanic ejecta from the Toba caldera; Possible relevance to Cretaceous/Tertiary boundary phenomena: Geology, v. 14, p. 380–383.

Carter, N. L., Officer, C. B., and Drake, C. L., 1989, Dynamic deformation of quartz and feldspar; Clues to causes of some natural crises: Tectonophysics (in press).

Cisowski, S. M., 1988a, Paleomagnetism of Manson structure cores inconsistent with a K/T link: 19th Lunar and Planetary Sciences Conference: Houston, Texas, Lunar and Planetary Institute, p. 188.

—— , 1988b, Analogues for magnetic microspherules associated with the K/T and upper Eocene extinction events; 19th Lunar and Planetary Sciences Conference: Houston, Texas, Lunar and Planetary Institute, p. 186.

Clark, D., Cheng-Yuan, W., Orth, C., and Gilmore, J., 1986, Conodont survival and low iridium abundances across the Permian–Triassic boundary in south China: Science, v. 233, p. 984–986.

Clark, R. P., and Edholm, O. G., 1985, Man and his thermal environment: London, Butler and Tanner, Ltd., 253 p.

Cook, M. A., 1968, The science of high explosives: New York, Reinhold Book Corp., 440 p.

Courtillot, V., Besse, J., Vandamme, D., Montigny, R., Jaegher, J-J., and Cappetta, 1986, Deccan flood basalts at the Cretaceous/Tertiary boundary: Earth and Planetary Science Letters, v. 80, p. 361–374.

Crocket, J. H., Officer, C. B., Johnson, G. D., and Wezel, F. C., 1986, Distribution of noble metals, arsenic across the Cretaceous/Tertiary boundary at Gubbio, Italy; Iridium variation as a constraint on the duration and nature of the Cretaceous/Tertiary boundary events: Geology, v. 16, p. 77–80.

Cronin, T. M., 1985, Speciation and stasis in marine ostracoda; Climatic model of evolution: Science, v. 227, p. 60–62.

Dauphin, J. P., 1983, Eolian quartz granulometry as a paleowind indicator in the northeast equatorial Atlantic, North Pacific and southeast equatorial Pacific [Ph.D. thesis]: Kingston, University of Rhode Island, 335 p.

Davies, G. J., 1973, Solidification and casting: New York, John Wiley and Sons, 205 p.

de Silva, S. L., and Sharpton, V. L., 1988, The K-T boundary debate; A volcanological perspective; 19th Lunar and Planetary Sciences Conference: Houston, Texas, Lunar and Planetary Institute, p. 273–274.

Douglas, M. M., 1981, Thermoregulatory significance of thoracic lobes in the evolution of insect wings: Science, v. 211, p. 84–85.

Duce, R. A., 1986, Aeolian mineral particles; Effects of atmospheric and marine processes: EOS Transactions of the American Geophysical Union, v. 44, p. 898.

Dulleforce, T. A., Buchanan, D. J., and Peckover, R. S., 1976, Self-triggering of small-scale fuel-coolent interactions; 1, Experiments: Journal of Physics, D, Applied Physics, v. 9, p. 1295–1303.

Ekdale, A. A., and Bromley, R. G., 1984, Sedimentology and ichnology of the Cretaceous–Tertiary boundary in Denmark; Implications for causes of the terminal Cretaceous extinction: Journal of Sedimentary Petrology, v. 54, p. 681–703.

Entwisle, A. R., 1971, The kinetics of martensite formation in steel: Metallurgical Transactions, v. 2, p. 2395.

Erben, H. K., Ashraf, A. R., Krumsiek, K., and Thein, J., 1983, Some dinosaurs survived the Cretaceous "final event" [abs.]: Terra Cognita, v. 3, p. KA6.

Fassett, J., and Obradovich, J., 1986, A high precision ^{40}Ar-^{39}Ar age spectrum plateau for 74 Ma from the uppermost, upper Cretaceous Fruitland formation–Kirtland shale, identifies paleomagnetic chron 33-normal in the San Juan basin, New Mexico: Geological Society of America Abstracts with Programs, v. 18, p. 598.

Ferguson, M.W.J., and Joanen, T., 1982, Temperature of egg incubation determines sex in *Alligator mississipiensis*: Nature, v. 296, p. 850–853.

—— , 1983, Temperature-dependent sex determination in *Alligator mississipiensis*: Journal of Zoology, v. 200, p. 143–177.

Fischer, K. M., Creager, K. C., and Jordan, T. J., 1986, Mapping the Tonga Slab with residual-sphere analysis; 16th Conference on Mathematical Geophysics, Oosterback, The Netherlands, June 22–28: European Union of Geosciences, p. L10.

Fricke, A., and Schreyer, W., 1987, Further fluid inclusion studies on minerals from the Vredefort structure, and comparisons with shocked Sudbury rocks; International Workshop on Cryptoexplosions and Catastrophes in the Geological Record, 6-10 July, Payrs, South Africa: University of the Witwatersrand.

Gartner, S., and McGuirk, J. P., 1979, Terminal Cretaceous extinction scenario for a catastrophe: Science, v. 206, p. 1272.

Graup, G., 1987, A volcanic aerosol model for the Cretaceous–Tertiary events; Results from the Lattengebirge section, Bavarian Alps; International Workshop on Cryptoexplosions and Catastrophes in the Geologic Record, 6-10 July, Parys, South Africa: University of the Witwatersrand.

Haack, S. C., 1986, A thermal model of the Sailback Pelycosaur: Paleobiology, v. 12, p. 450–458.

Hallam, A., 1987, End-Cretaceous mass extinction event; Argument for terrestrial causation: Science, v. 238, p. 1237–1242.

Hanks, T. C., and Thatcher, W., 1972, A graphical representation of seismic source parameters: Journal of Geophysical Research, v. 77, p. 4393–4405.

Hays, J. D., and Pitman, W. C., III, 1973, Lithospheric plate motion, sea-level changes, and climatic and ecological consequences: Nature, v. 246, p. 18–22.

Herman, Y., 1986, Survival and extinction of marine biota at the end of the Mesozoic era: Geological Society of America Abstracts with Programs, v. 18, p. 635.

Hino, K., 1959, The theory and practice of blasting: Nippon Kayaku, Yamaguchi-Ken.

Hsü, K. J., 1984, Strangelove ocean at the beginning of the Tertiary; Chapman Conference on Natural Variations in Carbon Dioxide and the Carbon Cycle, Tarpon Springs Abstracts: American Geophysical Union.

Hut, P., Alvarez, W., Elder, W. P., Hansen, T., Kauffman, E. G., Keller, G., Shoemaker, E. M., and Weissman, 1987, Comet showers as a cause of mass extinctions: Nature, v. 329, p. 118–126.

Ishihara, K., 1985, Dynamical analysis of volcanic explosion: Journal of Geodynamics, v. 3, p. 327–349.

Jansa, L. F., and Pe-Piper, G., 1987, Identification of an underwater extraterrestrial impact crater: Nature, v. 327, p. 612–614.

Kanamori, H., 1972, Mechanism of tsunami earthquakes: Physics of the Earth and Planetary Interiors, v. 6, p. 346–359.

Kanamori, H., and Anderson, D. L., 1975, Theoretical basis of some empirical relations in seismology: Bulletin of the Seismology Society of America, v. 65, p. 1073–1095.

Kanamori, H., and Given, J. W., 1982, Analysis of long-period seismic waves excited by the May 18, 1980, eruption of Mount St. Helens; A terrestrial monopole?: Journal of Geophysical Research, v. 87, p. 5422–5432.

Kanamori, H., Given, J. W., and Lay, T., 1984, Analysis of seismic body waves excited by the Mount St. Helens eruption of May 18, 1980: Journal of Geophysical Research, v. 89, p. 1856–1866.

Kanamori, H., Ekstrom, G., Dziewonski, A., and Barker, J. S., 1986, An anomalous seismic event near Tori Shima Japan; A possible magma injection event: EOS Transactions of the American Geophysical Union, v. 67, p. 1117.

Kasahara, K., 1981, Earthquake mechanics: London, Cambridge University Press, 248 p.

Katz, D. L., and Sliepovich, C. M., 1971, LNG/water explosions; Cause and effect: Hydrocarbon Processing, p. 240–244.

Kauffman, E. G., 1982, Environmental deterioration and graded extinctions at the end of the Cretaceous: American Association for the Advancement of Science Abstracts 148th National Meeting, Washington, D.C., p. 48.

Keefer, D. K., 1984, Rock avalanches caused by earthquakes; Source characteristics: Science, v. 233, p. 1288–1299.

Keller, G., D'Hondt, S., and Vallier, T. L., 1983, Multiple microtektite horizons in Upper Eocene marine sediments; No evidence for mass extinctions: Science, v. 221, p. 150–152.

Kennett, J. P., 1988, Proceedings of the Ocean Drilling Program, Leg 113, Sites 689, 690 (in press).

Kerr, R. A., 1987, Asteroid impact gets more support: Science, v. 236, p. 666–668.

Klinger, L. F., 1988a, The bryophyte paludification hypothesis of peatland formation; Implications for climate change and mass extinctions; Chapman Conference on Gaia Hypothesis, San Diego, March 7–11, American Geophysical Union, p. 14.

—— , 1988b, Successional change in vegetation and soils of southeast Alaska [Ph.D. thesis]: Boulder, University of Colorado, 234 p.

Knight, C. A., 1967, The freezing of supercooled liguids: London, D. Van Nostrand Co., Inc., 145 p.

Koepnick, R. B., Burke, W. H., Denison, R. E., Hetherington, E. A., Nelson, H. F., Otto, J. B., and Waite, L. E., 1985, Construction of the seawater $^{87}Sr/^{86}Sr$ curve for the Cenozoic and Cretaceous; Supporting data: Chemical Geology, Isotope Geosciences Section, v. 58, p. 55–81.

Larson, D. B., 1977, The relationship of rock properties to explosive energy coupling: Livermore, California, Lawrence Livermore Laboratory Report UCRL 52204.

Lerbekmo, J. F., Sweet, A. R., and St. Louis, R. M., 1987, The relationship between the iridium anomaly and palynological floral events at three Cretaceous–Tertiary boundary localities in western Canada: Geological Society of America Bulletin, v. 99, p. 325–330.

Lewin, R., 1988, Egg-laying birds remains a hot issue: Science, v. 239, p. 465.

Lightfoot, P. C., Hawkesworth, C. J., and Sethna, S. F., 1987, Petrogenesis of rhyolites and trachytes from the Deccan trap; Sr, Nd, and Pb isotope and trace element evidence: Contributions to Mineralogy and Petrology, v. 95, p. 44–54.

Loper, D. E., and McCartney, K., 1988, Shocked quartz found at the K/T boundary; A possible endogenous mechanism; Eos Transactions of the American Geophysical Union, v. 69, p. 961.

Lovett, G. M., and Reiners, W. A., and Olson, R. K., 1982, Cloud droplet deposition in subalpine balsam fir forests; Hydrological and chemical input: Science, v. 218, p. 1303–1304.

MacDougall, J. D., 1988, Seawater strontium isotopes, acid rain, and the Cretaceous–Tertiary boundary: Science, v. 239, p. 485–487.

Margolis, S. V., Mount, J. F., Doehne, E., Showers, W., and Ward, P., 1987, The Cretaceous/Tertiary boundary carbon and oxygen isotope stratigraphy, diagenesis, and paleoceanography at Zumaya, Spain: Paleoceanography, v. 2, p. 361–377.

McLean, D. M., 1978, A terminal Mesozoic greenhouse; Lessons from the past: Science, v. 201, p. 401–406.

—— , 1981, A test of terminal Mesozoic "catastrophe": Earth and Planetary Science Letters, v. 53, p. 103–108.

—— , 1985a, Deccan traps mantle degassing in the terminal Cretaceous marine extinctions: Cretaceous Research, v. 6, p. 235–259.

—— , 1985b, Mantle degassing unification of the trans-K-T geobiological record: Evolutionary Biology, v. 19, p. 287–313.

—— , 1986, Embryogenesis dysfunction in the Pleistocene/Holocene transition; Mammalian extinctions, dwarfing, and skeletal abnormality, in McDonald, J. N., and Bird, S. O., eds., The Quaternary of Virginia: Virginia Division of Mineral Resources Publication 75, 137 p.

Medenbach, O., Fricke, A., and Schreyer, W., 1987, Fluid inclusions along shock-induced planar elements in minerals from the basement rocks of the Vredefort structure; Fingerprints of an endogenic origin?; International Workshop on Cryptoexplosions and Catastrophes in the Geological Record, 6–10 July, Parys, South Africa: University of the Witwatersrand.

Messier, R., Takamori, T., and Roy, R., 1975, Observations on the "explosive" crystallization of non-crystalline Ge: Solid State Communications, v. 16, p. 331–314.

Moore, J. G., and Rice, C. J., 1984, Chronology and character of the May 18, 1980, explosive eruptions of Mount St. Helens, in Explosive volcanism; Inception, evolution, and hazards: Washington, D.C., National Academy of Sciences Press.

Morey, G. W., 1922, The development of pressure in magmas as a result of crystallization: Washington Academy of Sciences Journal, v. 12, p. 219–230.

Mount, L. E., 1979, Adaptation to thermal environment: Baltimore, Maryland, University Park Press, 333 p.

Nellis, W. J., Ree, F. H., van Thiel, M., and Mitchell, A. C., 1981, Shock compression of liquid carbon monoxide and methane to 90 GPa (900 kbar): Journal of Chemical Physics, v. 75, p. 3055–3063.

Nicolaysen, L., 1985, Renewed ferment in the earth sciences, especially about power supplies for the core, for the mantle, and for crises in the faunal record: South African Journal of Science, v. 81, p. 120–132.

Norman, D., 1985, The illustrated encyclopedia of dinosaurs: New York, Crescent Books, 208 p.

Nriagu, J. O., and Holdway, D. A., and Coker, R. D., 1987, Biogenic sulphur and the acidity of rainfall in a remote area of Canada: Science, v. 238, p. 1189–1192.

Oelofsen, B. W., 1978, Atmospheric carbon dioxide/oxygen imbalance in the Late Cretaceous, hatching of eggs, and the extinction of biota: Palaeontologica Africana, v. 21, p. 41–45.

Officer, C. B., Hallam, A., Drake, C. L., and Devine, J. D., 1987, Late Cretaceous and paroxysmal Cretaceous/Tertiary extinctions: Nature, v. 236, p. 143–149.

O'Keefe, J. D., and Ahrens, T. J., 1988, Impact production of CO_2 by the K-T extinction bolide, and the resultant heating of the whole earth; 19th Lunar and Planetary Sciences Conference: Houston, Texas, Lunar and Planetary Institute, p. 885.

Olmez, I., Finnegan, D. L., and Zoller, W. H., 1986, Iridium emissions from Kilauea Volcano: Journal of Geophysical Research, v. 91, p. 653–663.

Orth, C. J., Quintana, L. R., Gilmore, J. S., Grayson, R. C., Jr., and Westergaard, E. H., 1986, Trace-element anomalies at the Mississippian/Pennsylvanian boundary in Oklahoma and Texas: Geology, v. 14, p. 986–990.

Owen, R. M., and Rea, D. K., 1985, Sea floor hydrothermal activity links climate to tectonics; The Eocene carbon dioxide greenhouse: Science, v. 227, p. 166–169.

Perche-Nielson, K., 1987, Calcareous, siliceous, and organic-walled microfossils at the K/T boundary; International Workshop on Cryptoexplosions and Catastrophes in the Geological Record, 6–10 July, Parys, South Africa: University of the Witwatersrand.

Perche-Nielson, K., McKenzie, J., and He, Q., 1982, Biostratigraphy and isotope stratigraphy and the "catastrophic" extinction of calcareous nannoplankton at the Cretaceous/Tertiary boundary, in Silver, L. T., and Schultz, P. H., eds., Geological implications of impacts of large asteroids and comets on the Earth: Geological Society of America Special Paper 190, p. 353–371.

Pilant, W. L., 1979, Elastic waves in the Earth: New York, Elsevier Scientific Publishing Co., 493 p.

Pitman, W. C., III, 1978, Relationship between eustacy and stratigraphic sequences of passive margins: Geological Society of America Bulletin, v. 89, p. 1389–1403.

Playford, P. E., McLaren, D. J., Orth, C. J., Gilmore, J. S., and Goodfellow, W. D., 1984, Iridium anomaly in the Upper Devonian of the Canning Basin, western Australia: Science, v. 226, p. 437–439.

Rampino, M. R., and Volk, T., 1988 DMS and the K/T Boundary; Phytoplankton extinction, reduction in cloud albedo, and sudden climate warming; Chapman Conference on Gaia Hypothesis, San Diego, March 7–11: American Geophysical Union, p. 23.

Raup, D. M., and Sepkoski, J. J., 1984, Proceedings of the National Academy of Science, USA: Washington, D.C., National Academy of Science, v. 81, p. 801.

Reed-Hill, R. E., 1978, Physical metallurgical principles: London, D. Van Nostrand Co., Inc., 320 p.

Rice, A., 1972, Some Benard convection experiments, their relationship to viscous dissipation, and possible periodicity in sea floor spreading: Journal of Geophysical Research, v. 77, p. 2514–2525.

——, 1982, Soret convection and rheology (viscous dissipation); Arguments for whole mantle convection: Physics of the Earth and Planetary Interiors, v. 29, p. 330–343.

——, 1985, The mechanism of the Mt. St. Helens eruption and speculations regarding Soret effects in planetary dynamics: Geophysical Surveys, v. 7, p. 303–384.

——, 1987, Shocked minerals at the K/T boundary; Explosive volcanism as a source: Physics of the Earth and Planetary Interiors, v. 48, p. 167–176.

——, 1988, Nonlinear rate equations describing exploding magma chambers; Inferences from solutions: Terra Cognita, v. 8, p. 121.

Rice, A., 1981, Convective fractionation; A mechanism to provide cryptic zoning (macrosegregation), layering, crescumulates, banded tuffs and explosive volcanism in igneous processes: Journal of Geophysical Research, v. 86, p. 405–417.

Rice, A., and Eichelberger, J., 1976, Convection in rhyolite magma: EOS Transactions of the American Geophysical Union, v. 57, p. 1024.

Rice, A., and Fairbridge, R., 1975, Thermal runaway in the mantle and neotectonics: Tectonophysics, v. 29, p. 59–72.

Rich, J. E., Johnson, G. L., Jones, J. E., and Campsie, J., 1986, A significant correlation between fluctuations in seafloor spreading rates and evolutionary pulsations: Paleoceanography, v. 1, p. 85–95.

Rigby, J. K., Jr., Rigby, J. K., Sr., and Sloan, R. E., 1986, The potential for an unconformity near the Cretaceous/Tertiary boundary, Basal Tullock Formation, McCone County, Montana: Geological Society of America Abstracts with Programs, v. 18, p. 730.

Rigby, J. K., Jr., Newman, K. R., Van Der Kaars, J.S.S., Sloan, R. E., and Rigby, J. K., 1987a, Dinosaurs from the Paleocene part of the Hell Creek Formation, McCone County, Montana: Research Letters of the Society of Economic Paleontologists and Mineralogists, p. 296–302.

Rigby, J. K., Jr., Rice, A., and Currie, P. J., 1987b, Dinosaur thermoregulatory Cretaceous/Tertiary survival strategies: Geological Society of America Abstracts with Programs, v. 19, p. 820.

Rodean, H. C., 1970, Explosion-produced ground motion; Technical summary with respect to seismic hazards: Symposium on Engineering with Nuclear Explosives, Las Vegas, Nevada, Jan. 14–16, American Nuclear Society/Atomic Energy Commission, p. 1024–1050.

Roth, P. C., Blanchard, C., Harte, J., Michaels, H., and El-Ashray, M. T., 1985, The America west's acid rain test: World Resource Institute Research Report no. 1, 50 p.

Rutherford, M. J., Sigurdsson, H., Carey, S., and Davis, A., 1985, The May 18, 1980, eruption of Mount St. Helens; 1, Melt composition and experimental phase equilibrium: Journal of Geophysical Research, v. 90, p. 2929–2947.

Sanford, R. L., Jr., Saldarriaga, J., Clark, K. E., Uhl, C., and Herrera, R., 1985, Amazon rain-forest fires: Science, v. 227, p. 53–55.

Schleip, E., Willnecker, R., Herlach, D. M., and Gorler, G. P., 1987, Measurement of ultra-rapid solidification rates in largely undercooled bulk melts by a high-speed photosensing device: Institut fur Raumsimulation, DFVLR, D-5000 Koln 90, FRG, preprint.

Sethna, S. F., and Battiwala, H. K., 1977, Chemical classification of the intermediate and acid rocks (Deccan trap) of Salsette Island, Bombay: Journal of the Geological Society of India, v. 18, p. 323–330.

Shklovskii, V. A., and Druinskii, E. I., 1986, Explosive nonisothermal growth of a spherical phase-transition center during the decay of frozen metastable states; Soviet physics: Journal of Experimental and Theoretical Physics, v. 63, p. 137–141.

Sloan, R. E., Rigby, J. K., Jr., Van Valen, L. M., and Gabriel, D., 1986, Gradual dinosaur extinction and simultaneous ungulate radiation in the Hell Creek Formation: Science, v. 232, p. 629–633.

Stigler, S. M., and Wagner, M. J., 1987, A substantial bias in nonparametric tests for periodicity in geophysical data: Science, v. 238, p. 940–945.

Stothers, R. B., and Rampino, M. R., 1983, Historic volcanism, European dry fogs, and Greenland acid precipitation, 1500 B.C. to A.D. 1500: Science, v. 222, p. 411–413.

Sukheswala, R. N., and Sethna, S. F., 1962, Deccan traps and associated rocks of the Bassein area: Journal of the Geological Society of India, v. 3, p. 125–146.

Thorarinsson, S., 1969, The Lakagigar eruption of 1783: Bulletin Volcanologique, v. 33, p. 919–929.

Teller, E., Talley, W. K., Higgins, G. H., and Johnson, G. W., 1968, The constructive uses of nuclear explosives: New York, McGraw-Hill Book Co., 320 p.

Tredoux, M., de Wit, M. J., Hart, R. J., Lindsay, N. M., and Sellschop, J.P.F., 1987, Chemostratigraphy across the Cretaceous–Tertiary boundary at localities in Denmark and New Zealand; A case for the terrestrial origin of the platinum group element anomaly; International Workshop on Cryptoexplosions and Catastrophes in the Geological Record, 6–10 July, Parys, South Africa: University of the Witwatersrand.

Uematsu, M., Duce, R. A., and Prospero, J. M., 1985, Deposition of atmospheric mineral particles in the North Pacific Ocean: Journal of Atmospheric Chemistry, v. 3, p. 123–138.

Utsu, T., and Seki, A., 1955, A relation between the area of aftershock region and the energy of main shock: Zisin, Journal of the Seismological Society of Japan, v. 7, p. 233–240.

Vandamme, D., Besse, J., Courtillot, V., Motigny, R., Jaeger, J-J. and Cappetta, H., 1986, Deccan flood basalts at the Cretaceous–Tertiary boundary?: EOS Transactions of the American Geophysical Union, p. 67.

Voight, B., 1982, Time scale for the first moments of the May18 eruption: U.S. Geological Survey Professional Paper 1250, p. 69–86.

Voight, B., Glicken, H., Janda, R. J., and Douglass, P. M., 1982, Catastrophic rockslide avalanche of May 18: U.S. Geological Survey Professional Paper 1250, p. 347–377.

Wilde, P., Berry, W.B.N., Quinby-Hunt, M. S., Orth, C. J., Quintana, L. R., and Gilmore, J. S., 1986, Iridium abundances across the Ordovician–Silurian stratotype: Science, v. 233, p. 339–341.

Williams, H., and McBirney, A. R., 1979, Volcanology: San Francisco, California, Freeman, Cooper and Co., 397 p.

Wilson, L., and Huang, T. C., 1979, The influence of shape on the atmospheric settling velocity of volcanic ash: Earth and Planetary Science Letters, v. 44, p. 311–324.

Wolbach, W. S., Lewis, R. S., and Anders, E., 1985, Cretaceous extinctions; Evidence for wildfires and search for meteoritic material: Science, v. 230, p. 167–170.

Wolfe, J. A., and Upchurch, G. R., Jr., 1986, Changes in climate and diversity patterns across the Cretaceous–Tertiary boundary in the Raton Basin, New Mexico and Colorado: Geological Society of America Abstracts with Programs, v. 18, p. 793.

Zinsmeister, W. J., Feldmann, R. M., Woodburn, M. O., Kooser, M. A., Askin, R. A., and Elliot, D. E., 1987, Faunal transitions across the K/T boundary in Antarctica: Geological Society of America Abstracts with Programs, v. 19, p. 906.

Zoller, W. H., Parrington, J. R. and Kotra, J.M.P., 1983, Iridium enrichment in airborne particles from Kilauea Volcano; January, 1983: Science, v. 222, p. 1118–1121.

MANUSCRIPT RECEIVED BY THE SOCIETY JUNE 21, 1989

NOTES ADDED IN PROOF

Since submission of this chapter, several new, important factors have arisen that need to be addressed.

McHone and others (1989) report the discovery of a stishovite at the K/T boundary; they allege that such features cannot be retained in a volcanic environment because of high temperatures and conclude that impact, not volcanism, is the cause of the K/T event. Within several meters of detonating material, the pressure starts to fall off as $1/r^2$, which implies that stishovite formation could extend approximately 500 m out from the wall of the magma chamber if 90 kbar is taken as the pressure of formation. This neglects factors unknown for stishovite formation: the effect of impurities, load rate, etc. These factors serve to lower the transition pressure in other materials. If stishovite behaves similarly, its region of formation could extend even farther from the magma chamber. Jointing and cracks serve as stress concentrators and could double such ranges of formation. McHone and others indicate, however, that stishovite can be destroyed by "temperatures as low as 300 °C." The greatest distance from a convecting magma chamber some 500 m across at which such temperatures could be attained during its cooling history is itself approximately 250 m (e.g., Jaeger, 1964), assuming no cooling from natural convection ground-water circulation set up by the heat of the magma chamber. There is definitely room for survivability of stishovite in the volcanic environment. Although shocked minerals have been reported for Mt. St Helens (see main text), there has been no search for stishovite. If a search for stishovite is undertaken it must be kept in mind that if such material comes from depths approaching 5 km, little may reach the surface. It has been reported that the explosion of Mt. St Helens may have taken place as deep as 7 km (see main text).

De Silva and Sharpton (1989, referenced in the text) have further argued that the work of Wilson (e.g., 1980) establishes that observed ballistics of volcanic discharge do not require high pressures. Wilson neglects the diffusion times necessary to exsolve volatiles from de-pressurizing magmas that would dictate months for significant degassing rather than the time taken to transit from the magma chamber to the surface, i.e., minutes from explosive ejecta. Since Wilson's mechanism is diffusion controlled, it simply cannot occur and has no relevance.

Boslough (1990) acknowledges that quench supersaturation explosions as suggested here are feasible but claims pressures are limited to less than 10 kbar. Boslough's conclusions neglect the caveat in the introduction of his own paper: the equation of state he employs is an approximation limited to pressures less than 10 kbar. The use of "explosive pressure" as described in the text yields much higher estimations. While Sandia Laboratories (e.g., Boslough) now accept quench supersaturation mechanisms as a viable deformation mechanism, Sharpton (personal communication, 1989) maintains the release of latent heat on solidification would slow the freezing process. This indicates more than confusion regarding nucleation and freezing temperatures, the release of latent heat simply driving the melt back up to the freezing point and providing activation energy to secure freezing (e.g., see Knight, 1967, in the references of the main text).

With respect to the nuclear winter arguments that attend the impact extinction theories, Rich and others (1988) report that dinosaurs that had become extinct in warmer climes survived in southern polar regions. Expanding in a condensed fashion on McLean's suggestion regarding the deleterious effect of excess ambient temperature on reproductive mechanisms, it is important to note that the mechanism of heat shock response is similar in bacteria, plants, and animals such as insects and amphibians; e.g., Drosophilia, yeast (Craig and others, 1982). Hence, this characteristic appears to cut across all evolutionary orders, including those that must have arisen earliest in the evolution of life. Heat shock in the laboratory typically consists of raising the temperature from say 23 to 36 °C or in extreme cases, as high as 37 to 42 °C for several hours (e.g., Finklestein and Strausbert, 1982; Pelham and Bienz, 1982). Although these temperatures are below the lethal temperature for a number of laboratory animals, they still effect significant changes in cell protein synthesis (e.g., Voellmy and Rungger, 1982). Organisms also respond by increasing the production of proteins that serve to repress heat shock. Particularly striking is the effect of excess heat on embryonic development. In Drosophilia, there is a significant reduction in neurons (e.g., Buzin and Bournais-Vardiabasis, 1982). Such effects carry over to rats. A single exposure to 40 to 41 °C for 40 to 60 minutes results in increased resorptions and abnormalities of the eyes, brain, and face of survivors (Reproductive Toxicology Center, 1983). Hyperthermia is known to be teratogenic ("monster" producing) in animals. Harvey and others (1981) provides a number of references. Hyperthermia is also a human teratogen and in general most deleterious to fetal development, in particular the development of the nervous system. If maternal body temperatures exceed normal by only 1.5 to 2.5 °C, embryonic development is severely impaired. The particular damage induced by hyperthermia is related to the specific stage of fetal development at which the several-degree temperature rise occurs. The list of abnormalities induced in human offspring by fetal hyperthermia is long: Exencephaly (brain forms outside the skull), anencephaly (the brain is missing), encephalocoele (hernia of brain through skull), holoprosencephaly (failure of closure of the forebrain), microencephaly (stunted brain), microthalmia (abnormally small eyes), talipes (club foot), arthrogryposis (persistent spasms), exomphalos, agnathia (absence of the lower jaw), reduced maxilla, facial clefts, hydrocephalus, cavities of the cerebrocortical white matter, catarcts, strabismus (squinting), blindness, thin sclera (eye container), hypertonus of the gastrocnius muscle, distortion of the tarsus (ankle), hyplasia (arrested development of toes or teeth), displacement of the tuber calcani (heel), fragile and bowed tibia, agenesis or hypoplasia of the fibula, distortion and immobility of the carpus (wrist), dysplasis (abnormal development of the spinal cord), renal agenesis (little or no kidneys and/or urinary tract), vertebral anomalies, Kyphosis (hunchback), lordosis (swayback), scoliosis, meroanencephaly, teratology of Fallot (congenital heart defects), anophthalmia (missing or vestigal eyes), coloboma (absence of ocular tissue), placental infarcts, necrosis of the decidua basalts (base tissue for the placenta), meningeal and subcutaneous hemorrhages, Hirschprung's disease (total or partial lack of the altimentary tract nervous system), hypospadias (urethra opens under penis), spina bifida, sensorineuronal deafness, retardation, nystagmus (involuntary jerking of the eye), Down's syndrome, cerebral palsy, Oesophageal Artesia (missing orifice), tracheo-oesophageal fistula (abnormal passage), urethral stricture and blockage, hydronephrosis (kidney degeneration), coarctation (constriction) of aorta, artresia of the aortic valve, transposition of the great vessels, interventrical septal defects, patent ductus arteriosus (hole in the heart), pulmonary, aortic stenosis (constriction), enlarged hearts, systolic murmors, bifid uvula, bilateral athetosis, spastic diplegia, pyloric stenosis, umbilical hernia, inguinal hernia, increased miscarriages, increased stillborn births, increased resorptions, naevi, etc. (e.g., Lipson, 1987; Edwards, 1979; Edwards, 1986; McDonald, 1961).

In laboratory as well as domestic and wild animals, most of the above defects appear on demand by application of a several-degree temperature rise at the appropriate stage of gestation (e.g., in monkeys, mice, rats, guinea pigs, hamsters, sheep, chickens, farm animals, birds, etc.). It is known that cells in mitosis are the most sensitive to hyperthermia, showing loss of metaphase chromosome orientation, chromosome clumping, and disruption of the mitotic apparatus. For that matter, application of temperatures of 42 °C for one day causes heavy to total mortality of chickens in their first six days of incubation. Abnormalities in embryos were engendered with application of 43 °C for 7.5 minutes. Applied for 30 minutes, this temperature yields many necrotic cells in neuroepithelia. An application of 42 °C (one degree less than above) for 10 minutes led only to a reduction in protien synthesis (dwarfing?). Offspring of heated mothers were in general clumsy, slow, failed to bond with the mother, and died within hours or days after birth. Embryos cultured at 40.5 °C show good tissue organization, but large numbers of pyknotic nuclei are found in the brain and neural tube, with large deficits of neurons. Cell damage can be seen immediately on application of temperature.

It has been found that a 1 °C elevation in maternal temperature is equivalent to 50 rads of gamma radiation administered to the embryo during the most sensitive stage of gestation. This is considered to be equivalent to the radiation doses that led to birth defects from the atom bomb dropped at Hiroshima. It is now standard practice to warn pregnant women not to enter hot tubs or saunas for fear of damage to fetal development (e.g., Harvey and others, 1981). It is well known that excess temperature also impedes sperm production in mammals. Sparrows (more direct descendants of dinosaurs) have similar problems. Spermatogenesis occurs only at night when their body temperature is lowest (e.g., Edwards, 1986). European cattle gestate during the cool winter months. Bakker (personal communication) reports that generally only animals of mass less than 10 kg or less than one meter in length survived the early Tertiary.

In all creatures the embryonic central nervous system contains the cells with the greatest sensitivity to heat-shock insult. This sensitivity carries over to the mature animal. Prairie rattlers without protection from the midday sun will suffer brain death (as do other herptiles) and present no small hazard to any intrusion into their shade. Although reptiles do control temperature throughout, nowhere is temperature so carefully regulated as in the head. This applies to snakes, turtles, and lizards, as well as crocodilians, for which temperatures of 38 to 39 °C are lethal (e.g., Johnson, 1974).

Many mammals have aerated heads, which are thought to serve to cool the blood entering the brain. Cooling strategies in dogs and birds, i.e., panting, are thought to serve primarily to lower head temperatures: this by breathing in through the mouth and out the nose to cool the brain. Heat stroke in animals due to excessive exercise impacts the brain before any other organ. This hazard is particularly severe for some types of predators in tropical climes such as Africa. Cheetahs carefully pick microclimates to minimize this hazard (Bakker, personal communication, 1988). Raising the temperature of the human brain from a normal 37 to 42 °C (e.g., through sunstroke) will result in death. On exposure to the summer sun, these temperatures are reached in a remarkably short time for people with defective sweat glands (e.g., Cannon, 1980). Another tropical animal, the vampire bat, will perish if its temperature reaches 33 °C (Lyman and Winsatt, 1961). These points will be discussed in detail elsewhere (Rice, A., and Rigby, J. K., Jr., in preparation).

ACKNOWLEDGMENTS

I thank M. Lockley, R. Bakker, R. Parrish, S. Chatterjee, R. Lehman, E. Bray, B. Sayles, G. Burchfield, and a host of others for their useful input.

ADDITIONAL REFERENCES CITED

Boslough, M. B., 1990, A thermochemical model for shock-induced chemical reactions in porous solids; Analogs and contrasts to detonation: Proceedings of the 9th Symposium on Detonation (in press).

Buzin, C. H., and Bournais-Vardiabasis, N., 1982, The induction of a subset of heat-shock proteins by drugs that inhibit differentiation in Drosophila embryonic cell cultures, *in* Schlesinger, M. J., Ashburner, M., and Tissieres, A., eds., Heat shock from bacteria to man: New York, Cold Spring Harbor Laboratory, p. 387–394.

Cannon, W. B., 1980, The constancy of body temperature, *in* Satinoff, E., ed., Thermoregulation: Stroudsburg, Pennsylvania, Dowden, Hutchinson, and Ross, Inc., p. 14–43.

Craig, E., Ingolia, T., Slater, M., Manseau, L., and Bardwell, J., 1982, Drosophilia, yeast, and E. Coli genes related to the Drosophilia heat-shock genes, *in* Schlesinger, M. J., Ashburner, M., and Tisseres, A., Heat shock from bacteria to man: New York, Cold Spring Harbor Laboratory, p. 11–18.

Edwards, M. J., 1979, Is hyperthermia a human teratogen?: American Health Journal, v. 98, p. 277–280.

—— , 1986, Hyperthermia as a teratogen; A review of experimental studies and their clinical significances, *in* Teratogenesis, carcinogenesis, and mutagenesis, vol. 6: p. 563–582.

Harvey, M.A.S., McRorie, M. M., and Smith, D. W., 1981, Suggested limits to the use of the hot tub and sauna by pregnant women: CMA Journal, v. 125, p. 50–53.

Jaeger, J. C., 1964, Thermal effects of intrusions: Reviews of Geophysics, v. 2, p. 443–465 (see in particular the section on convecting magmas).

Johnson, C. R., 1974, Thermoregulation in crocodilians; 1, Head-body temperature control in the Papuan–New Guinean Crocodiles, Crocodylus Novaeguineae and Crocodylus Porosus: Comp. Biochem. Physiol., v. 49, p. 3–28.

Lipson, A., 1987, Hyperthermia and hirschsprung disease: Sydney, Australia, Children's Hospital Birth Defects Unit, FRACP Head (preprint).

Lyman, C. P., and Winsatt, W. A., 1961, Temperature regulation in the vampire bat, Desmodus Rotundus: Physiological Zoology, v. 34, p. 101–109.

McDonald, A. D., 1961, Maternal health in early pregnancy and congenital defect: British Journal of Preventative and Social Medicine, v. 15, p. 154–166.

McHone, J. F., Nieman, R. A., Lewis, C. F., and Yates, A. M., 1989, Stishovite at the Cretaceous-Tertiary boundary, Raton, New Mexico: Science, v. 243, p. 1182–1184.

Pelham, H., and Bienz, M., 1982, DNA sequences required for transcriptional regulation of the Drosphilia hsp 70 heat shock gene in monkey cells and Xenopus Oocytes, *in* Schlesinger, M. J., Ashburner, M., and Tissieres, A., Heat, eds., Heat shock from bacteria to man: New York, Cold Spring Harbor Laboratory, p. 43–48.

Rich, P. V., and 6 others, 1988, Evidence for low temperatures and biological diversity in Cretaceous high latitudes of Australia: Science, v. 242, p. 1403–1406.

Voellmy, R., and Rungger, D., 1982, Heat-induced transcription of Drosophilia heat shock genes in Xenopus Oocytes, *in* Schlesinger, M. J., Ashburner, M., and Tissieres, A., eds., Heat shock from bacteria to man: New York, Cold Spring Harbor Laboratory, p. 49–56.

Wilson, L., 1980, Relationships between pressure, volatile content, and ejecta velocity in three types of volcanic explosion: Journal of Volcanology and Geothermal Research, v. 8, p. 297–313.

# Plant successions and interruptions in Miocene volcanic deposits, Pacific Northwest

Ralph E. Taggart and Aureal T. Cross
Department of Geological Sciences, Department of Botany and Plant Pathology, Michigan State University, East Lansing, Michigan 48824

ABSTRACT

Analyses of pollen and spore distribution in samples taken in close stratigraphic succession from several sections of the Neogene Succor Creek, Stinking Water, and Trapper Creek floras indicate that the source-vegetation mosaic at each site was in a state of ecological disequilibrium. All these floras are preserved in volcaniclastic sediments, and in each case the presence of fossiliferous strata is closely linked in both space and time with local eruptive activity. The exposed stratigraphic sections appear to represent relatively short periods of time, probably less than 10,000 yr, excluding unrecognized diastems. In all cases, most of the time represented in the studied sections records successional vegetation rather than a mosaic of community types in equilibrium with the prevailing paleoclimate. In most cases, the ecological dynamics can be related to the disturbance or destruction of forests by local volcanic activity and/or fire, followed by distinctive seral stages, barring further disturbance, leading to the reestablishment of a diverse forest-community mosaic. Many of the megafossil collections from these floras have lacked sufficient stratigraphic control, and most represent a single seral stage or a mélange of seral stages. Reconstructing the megafossil assemblages that correspond to specific stages in the recovery continuum is rarely possible.

Pollen spectra characteristic of minimally disturbed communities are rare. Of the three floras studied, the Stinking Water represents the lowest paleoelevation and/or highest paleotemperature of forests dominated by broadleaved taxa throughout the area. Conifer stands were rare and confined to local topographic highs. The Succor Creek area supported a diverse array of broadleaved-dominated forests in the lowlands and conifer-dominated forests on adjacent low slopes, suggesting a cool but equable paleoclimate. The southern part of the Succor Creek area was probably higher and topographically more diverse than the northern part of the area. The Trapper Creek region was higher and/or cooler than the other sites and was dominated by coniferous forest. Broadleaved-dominated communities in the Trapper Creek area were significant only in the recovery intervals following disturbance.

INTRODUCTION

The large number of Neogene fossil-plant assemblages in the Pacific Northwest is a consequence of extensive episodic volcanism in the region. Throughout much of the history of terrestrial plants, the fossil record is strongly biased toward coastal plains and continental margins. Such sites offer a consistent combination of factors, including large volumes of water at moderate to low energies, immense loads of fine clastic sediments, a great number of plant parts derived from large watersheds, extensive areas of floodplain and interfluve, a large number of lakes, and the potential for the development of extensive biogenic swamps, all enhancing the potential for preservation of fossil plants. In contrast, inland/upland areas are comparatively poorly repre-

Taggart, R. E., and Cross, A. T., 1990, Plant successions and interruptions in Miocene volcanic deposits, Pacific Northwest, *in* Lockley, M. G., and Rice, A., eds., Volcanism and fossil biotas: Boulder, Colorado, Geological Society of America, Special Paper 244.

sented. Source watersheds tend to be small and lakes relatively less common. Streams and rivers tend to have higher energy levels, clastic sediments are generally coarser, floodplains have less area and are more subject to destruction, and swamps are uncommon and tend to be small when they occur.

The Neogene volcanism of the Pacific Northwest, in contrast to the "normal" clastic regime, facilitated the preservation of fossil assemblages over a wide geographic area and from a wide range of elevations. A number of aspects of local volcanism interact to facilitate preservation of plant material. First, lava flows, lahars (mudflows), and earthquakes can dam streams, creating ponds and lakes that were not characteristic of the pre-volcanic landscape. An example of this phenomenon is the Florissant flora of Colorado where lahars dammed a stream, creating Lake Florissant, which served as the basin of deposition for that Oligocene flora. Water accumulating in collapsed calderas also represents "new" lakes. The Oligocene Creede flora of Colorado and the Miocene Trout Creek flora of Oregon are both preserved in caldera lakes. In these cases and many others, the fossil record for specific sites is confined to such volcanogenic lakes and ends with the cessation of lake deposition.

Perhaps the most obvious effect of local volcanic activity in facilitating fossil preservation is the immense quantity of relatively fine volcanic ash produced by proximal volcanic centers. These volcanogenic sediments supplement the often-meager supply of normal clastics both by direct airfall into bodies of water and, perhaps more significantly, by reworking of poorly consolidated ash initially deposited on surrounding watersheds. The result is varved and deltaic deposits in lakes and a wide range of fluvial deposits producing volcanic analogs of sandstones, siltstones, shales, and mudstones. The majority of the megafossil assemblages in this region are preserved in such deposits. In addition to the direct input of ash into depositional sites, the drainage of water from ash-covered landscapes and the percolation of water through tuffaceous deposits greatly enrich these waters in dissolved silica and other elements normally present in limiting concentrations, fostering blooms of diatoms, which in turn can result in the preservation of plant materials in diatomites such as those at Trout Creek. Newly impounded waters, enriched in silica, typically flood preexisting forests, facilitating permineralization of wood. As a consequence, fossil stump fields and other fossil wood deposits are comparatively common throughout this region, although they are poorly studied.

This collection of inland and upland megafossil assemblages appears to offer excellent potential for ecological reconstruction. The floras themselves have been the subject of variable, but generally extensive, taxonomic study, and the composite flora is quite diverse, particularly with regard to trees and shrubs. Our own computer data base includes more than 350 published binomials and is far from complete. Although systematic interpretations are subject to continuing evaluation, many of these woody taxa appear, on morphological grounds, to be similar to specific extant species or lineages within extant genera. These similarities provide, in fact, the basis for what might be termed the "uniformitarian ecological" approach to ecological reconstruction. This approach, which was broadly applied by Chaney (1925) and Chaney and Axelrod (1959), infers the ecological characteristics of the fossil taxa on the basis of the known or presumed attributes of the presumed "equivalent" extant species. Axelrod has refined this approach (Axelrod, 1966; Axelrod and Bailey, 1969), reconstructing both geographic and elevational distribution of major forest types. Such a mode of interpretation obviously is dependent on two crucial factors: (1) the degree of "equivalence" between fossil and extant taxa, a matter of critical systematic judgement with obvious potential for varying degrees of error; and (2) the respect to which the ecological characteristics of individual taxa exhibit varying degrees of "constancy" in the temporal transition from the Neogene to the present. If specific taxa play a critical role in the ecological interpretation of a fossil assemblage, the importance of correct identification is obvious. The wholesale reinterpretation of Oligocene and Miocene floras, necessitated by the recognition of *Metasequoia* in contrast to the earlier "redwood forest" (*Sequoia*) model of Chaney (1948, 1951), is an excellent example of this problem. Ecological constancy is also problematic, at least in some lineages. Wolfe (1979) has pointed out that Neogene *Glyptostrobus* occurrences are invariably associated with forests of temperate aspect, while the extant *Glyptostrobus pensilis* of China occurs in association with paratropical rainforest.

Although many of the problems of identification and ecological constancy can be mitigated by concentrating on whole assemblages of plants, rather than by relying excessively on a smaller suite of ecological "indicators," MacGinitie (1941) has noted that the problem of circularity exists in that ecological perceptions can bias critical floristic assessments. Many workers would contend that the floristically based methods involve too many theoretical and practical problems to serve as the primary tool for paleoecological analysis. Wolfe (1971, 1978) attempted to avoid such problems by basing interpretations on leaf physiognomy. Such an approach is essentially taxon-free and is based on the quantitative occurrence of various attributes of leaf physiognomy (leaf size, margin, texture, etc.) in fossil suites, compared with modern vegetation.

In general, workers applying either floristic or physiognomic methodology in paleoecological reconstruction have accepted the validity of the fossil record itself and have concentrated their debate on the mode of interpretation. Our work has involved the integration of megafossil and palynological data in both short and long stratigraphic sequences of sediments from a number of Neogene localities in the Pacific Northwest region. Our studies suggest that there are a number of problems presented by plant assemblages in such sections that must be addressed prior to any attempt at paleoecological reconstruction, regardless of the methodology employed. One of the major problems concerns the extent to which the source vegetation in these floras was disturbed by the same local volcanic activity that was responsible for the preservation of the fossil flora. Our work clearly indicates that, in the rock record, much of the time represented by these

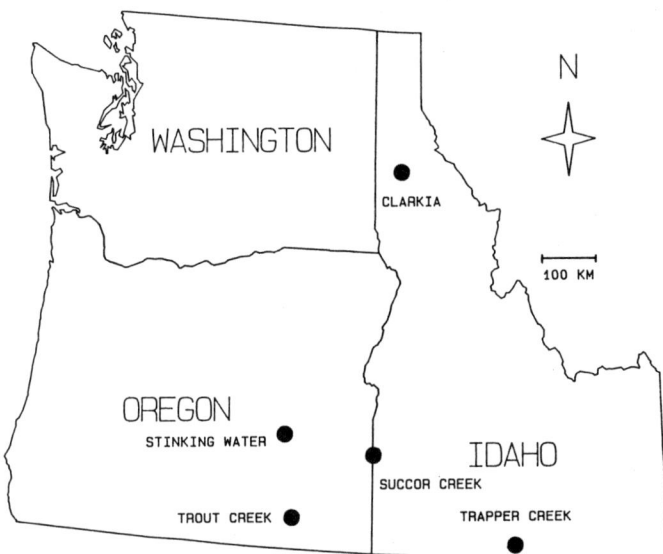

Figure 1. A regional map of the Pacific Northwest showing the distribution of the Neogene floras noted in the text.

fossil deposits, records vegetation in various stages of recovery in response to a wide variety of disturbance factors and mechanisms. In effect, the record appears dominated by vegetation in a state of ecological disequilibrium, complicating the task of reconstructing the relation between the fossil source vegetation and the prevailing paleoclimate.

THE FLORAS

Integration of megafossil data and stratigraphically controlled palynological data for three Neogene floras from the Pacific Northwest region will be used to demonstrate the scope of the problems that must be addressed. The Succor Creek flora of the Oregon-Idaho border region (Fig. 1) has been the most intensively studied. The most recent floristic monograph for this flora was undertaken by Graham (1965); this study also includes quantitative megafossil data as well as pollen and spore counts from three fossil localities. Later work on this assemblage has concentrated largely on stratigraphically controlled pollen and spore studies (Taggart, 1971, 1973, 1987; Taggart and Cross, 1974, 1980; Taggart and others, 1982; Cross and Taggart, 1982; Satchell, 1984). The megafossil record for the Stinking Water flora of eastern Oregon was summarized by Chaney and Axelrod (1959). The definitive monograph on the qualitative and quantitative megafossil composition of the Trapper Creek flora of southern Idaho was prepared by Axelrod (1964).

In terms of diversity, all three assemblages are dominated by broadleaved woody taxa, most of which were deciduous. Such physiognomic characterizations are of limited value, however, since a significant majority of the known regional floristic inventory (approximately 82 percent) consists of broadleaved taxa.

The quantitative representation of dominant genera for individual localities in these floras is plotted in Figure 2. Although most localities are obviously oak (*Quercus*) dominated, there is considerable variability in representation of both oaks and other taxa. The oaks have been subdivided into three broad categories. Lobed oaks (*Q. pseudolyrata* and *Q. prelobata*), which were undoubtedly deciduous in habit, are significant only at Stinking Water; there they dominate at locality P4120 but make up less than 25 percent of the oaks at locality P4006. A second group of presumably evergreen oaks (*Q. hannibali* and *Q. dayana*) are the major dominants at most Succor Creek localities and constitute over 50 percent of the oak material at P4006 at Stinking Water. The third oak element is represented by material assigned to *Q. simulata* (including *Q. consimilis* in Graham, 1965). Chaney and Axelrod (1959) allied this taxon with evergreen cyclobalanoid oaks, but Axelrod (1984) suggested that the taxon represents an extinct lineage of deciduous protobalanoid oaks. Wolfe (personal communication, 1988) suggests that *Q. simulata* is probably a sister group to the deciduous *Q. sadleriana,* noting that the leaves lack the coriaceous texture typical of evergreen oak leaves. In any case, this problematic oak is invariably present at Succor Creek localities and dominates at Graham's (1965) locality 2. It makes up about 25 percent of the oak material at Stinking Water P4006 and dominates the entire megafossil assemblage at Trapper Creek. The Trapper Creek assemblage is also unique in terms of the strong representation of montane conifers such as spruce (*Picea*), fir (*Abies*), and Douglas fir (*Pseudotsuga*). Although spruce is present at both Stinking Water and Succor Creek, and material of both fir and Douglas fir have been collected at Succor Creek, these taxa are quantitatively insignificant, as are pines (*Pinus*). Such patterns led Axelrod (1968) to reconstruct the Succor Creek and Stinking Water floras as relatively low-elevation broadleaved slope forest, while the Trapper Creek was considered to represent relatively more-upland conifer-hardwood forest.

While such quantitative data, representing tabulations of thousands of specimens, may appear to represent a sound basis for ecological reconstruction, this is far from the case. Tabulations based on previous collections have the potential for serious distortion due to "collection bias," a term referring to the natural tendency to collect and curate rare or particularly well-preserved material at the expense of relatively common taxa or material in fragmentary condition. It is possible to tabulate material on the outcrop in an attempt to counter this problem, but this simply creates new difficulties in making critical taxonomic determinations involving poorly preserved or fragmentary specimens. Fundamental errors in identification can adversely affect the tabulations and can rarely be corrected after the fact. A third alternative, returning all fossil material for study in the laboratory, is rarely practical given the resources available.

The practical and theoretical problems of such quantitative tabulations aside, the degree to which such quantitative suites represent the source vegetation of a paleowatershed is problematic. Modern taphonomic studies (Spicer, 1981; Spicer and Wolfe,

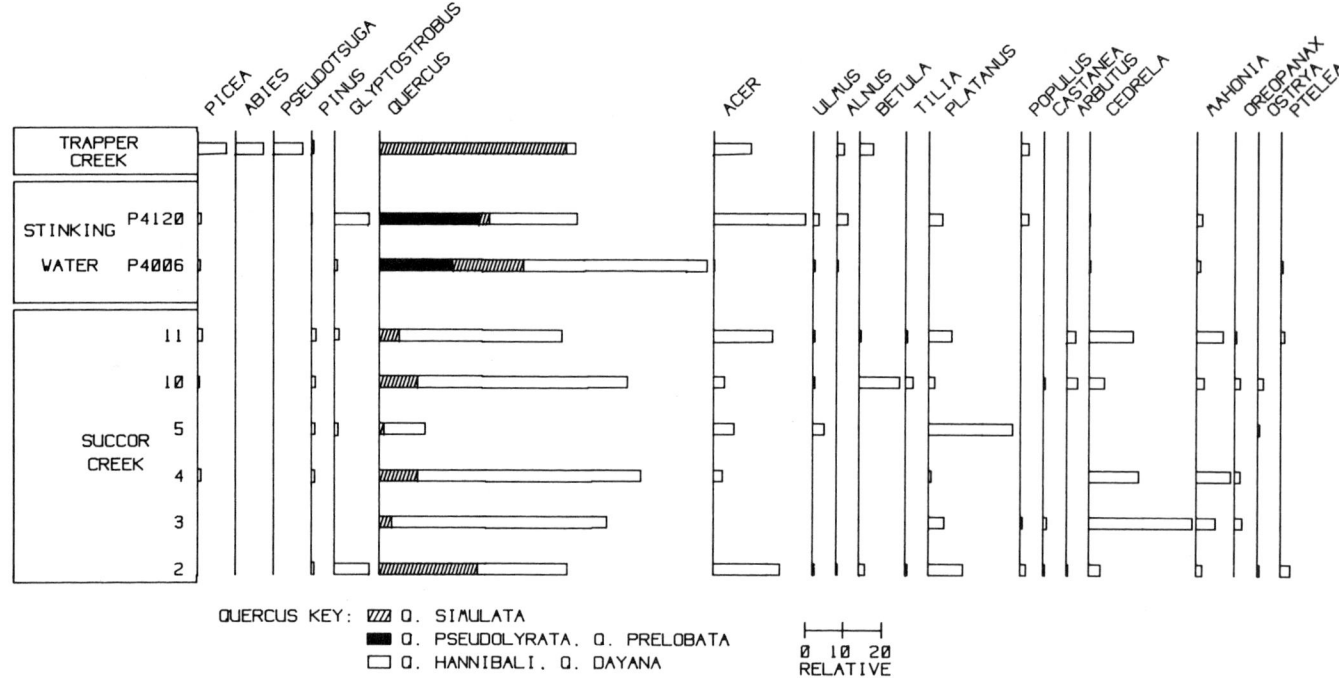

Figure 2. Quantitative representation of woody taxa at Trapper Creek (Axelrod, 1964), two localities in the Stinking Water (Chaney and Axelrod, 1959) flora, and six localities (2, 3, 4, 5, 10, 11) of the Succor Creek (Graham, 1965) flora. Only those localities with more than 100 specimens have been tabulated and only taxa representing at least 1 percent relative frequency at a minimum of one locality have been plotted.

1987) suggest that assemblages of plant megaremains accurately record the presence of taxa in the source vegetation, but that there is little correspondence in terms of quantitative relations. The effective "sampling radius" represented by larger plant parts can be quite short. Axelrod (1964) suggests that *Quercus simulata*, the oak making up more than 50 percent of the megafossil remains at Trapper Creek, was confined to lakeshore/streamside habitats and was not a significant component in nearby slope forests. To some extent, stratigraphically controlled pollen and spore data may serve as an alternative test on the nature of the source vegetation. Although the taxonomic resolution obtainable with pollen and spores is typically less than that achievable with megafossil material such as leaves, it is generally recognized that the potential for dispersal of pollen and spores is inherently greater than that for larger plant parts. Pollen and spore representation in fossil deposits is controlled by a network of taphonomic biases involving transport vectors, aerodynamic and hydrodynamic qualities, entomophily as opposed to anemophily, and so on. It can be reasonably assumed, however, that such biases are different than those operating on leaves and other large plant parts. Thus, pollen and spore representation in fossil suites can provide an independent check on the sampling fidelity of the megafossil record.

While the "sampling radius" for pollen suites is almost certainly greater than that of leaves, it is far short of providing a "regional" perspective on source vegetation. Modern pollen rain studies uniformly suggest that the dominant elements of pollen spectra, derived from forest vegetation, document pollen produced by proximal plant communities. Our own analysis of conifer and broadleaved pollen representation in transects from Mt. Mitchell in North Carolina (Cross and Taggart, 1982) indicates that the pollen spectra are quite sensitive to the composition of vegetation in the immediate area of the collection site.

The problems associated with inadequate stratigraphic control in the study of fossil assemblages will be a major focus in the discussions that follow.

ECOLOGICAL DYNAMICS IN SHORT STRATIGRAPHIC SEQUENCES

The vegetation mosaic on any landscape is always in a state of dynamic equilibrium under the influence of a wide range of biotic and abiotic factors. The nature of the mosaic and its relation to a specific landscape at any moment in time is a product of many factors, including the available biotic inventory (a product of immigration, emigration, extinction, and evolution), site factors (including substrate, drainage patterns, slope exposure, topography, etc.), climate, tolerance ranges and competitive status of the constituent taxa, and the consequences of the previous ecologic history of the specific site. In neoecology there is no difficulty in differentiating spatial and temporal heterogeneity; the time frame is fixed, and the spatial dimension is self-evident. In contrast, while the spatial dimension in the paleogeographic distribution of fossil localities is usually evident, time resolution is variable and must be considered a primary element in any analysis.

Many of the Neogene floras of the Pacific Northwest represent more than one locality. Some of these localities may be separated by many miles but are included in the context of a single flora on the basis of overall similarity in floristic composition, lithology, etc. While the material from these disparate localities is rightfully tabulated separately, the data from all available localities is typically pooled in the final analysis. Such compositing increases the diversity of the "flora" and is generally desirable in the sense that it improves our understanding of the biotic inventory of a specific region within the context of a formation or other limited stratigraphic unit where evolutionary change is not a major consideration. Such compositing is completely inappropriate in terms of paleoecological analysis, for it has the potential to obscure time-dependent factors that may be critically important in ecological reconstruction.

Figure 2 indicates the degree of quantitative variability between megafossil localities of the Succor Creek, Stinking Water, and Trapper Creek floras. Typically the fossiliferous exposures represented by such localities represent relatively short stratigraphic sequences, and their precise relation to the stratigraphy of other sites is not known. If all localities of a specific "flora" were strictly contemporaneous, the variability between localities would reflect the heterogeneity of the source-vegetation mosaic (Spicer and Wolfe, 1987). This, of course, assumes that similar taphonomic biases are operative in each case, something that may or may not be so. Since contemporaneity cannot be assumed, such variability may also reflect dynamic changes in the composition and distribution of components of the source-vegetation mosaic with time. Pooling data from multiple localities in a paleoecological analysis has the unfortunate consequence of effectively assuming contemporaneity without clearly stating the assumption.

Documentation of the range of dynamics of the source-vegetation mosaic with time requires a degree of stratigraphic control that has generally been lacking in most of the "classic" studies of the Neogene vegetation of the Pacific Northwest. Notable exceptions can be found in Graham's (1965) analysis of megafossil distribution through the composite Trout Creek section and Smiley and Rember's (1985) analysis of the Clarkia (northern Idaho) deposit. In both cases, successional dynamics are cited as the causative agent behind the considerable degree of variability that was noted.

We have previously noted the various ways in which the Neogene volcanism of the region resulted in preservation of fossil assemblages that would not have been preserved under conditions of "normal" clastic sedimentary processes. Volcanic activity at any specific site will be episodic. The appearance of a new volcanic center will result in a finite span of time characterized by intermittent moments of activity, followed by periods of quiescence of the center. Preservation of fossil assemblages will be tightly linked in temporal terms with periods of activity. The creation of impoundments is basically ephemeral because the newly created depositional basins may be drained by erosion and breaching of containment features, or infilling by sediments of the water body itself. The supply of a significant amount of airfall ash is obviously dependent on local eruptive activity and, while the supply of reworked, water-transported ash can obviously extend beyond the period of local activity, the supply of such material is finite, and mobilization of such material cannot continue indefinitely. Thus, the potential for preservation of a fossil assemblage becomes high with the activation of a local volcanic center and declines following a reduction in activity. Any given volcanic center will have a unique pattern of activity, but in general, the pattern will probably involve clusters of eruptive activity, separated by longer periods of quiescence. The pattern of volcanoclastic deposits, including fossil floras, will reflect such temporal clustering of eruptive episodes. One of the major difficulties in working with the Neogene assemblages in this region is the limited nature of typical stratigraphic exposures and the correlation of different exposures with little or no stratigraphic continuity. While such difficulties are commonly exacerbated by the fact that volcanoclastic exposures are easily eroded, they are basically a reflection of the geologically short duration of activity in the case of specific volcanic centers. McLeroy and Anderson (1966) determined, on sedimentological grounds, that the Oligocene lake sediments at Florissant represented perhaps 5,000 yr of deposition. Cross and Taggart (1982), in their discussion of the time represented by exposed sections in the Succor Creek area, suggested that most sections in the area represented no more than 10,000 yr, despite the fact that intermittent volcanic activity in the entire Succor Creek area may have spanned as much as 2 m.y.

The fact that a fossil deposit may represent a short temporal "window" complicates the ecological analysis of these Neogene assemblages precisely because such short sedimentary intervals are closely associated, temporally, with local volcanic activity. Climatic interpretations on the basis of paleovegetation data are based on the fact that climate is considered the major determinant of the vegetation developed on a site. Because the source-vegetation mosaic in a given area is in a state of dynamic equilibrium, it is assumed that the nature of the mosaic will "track" climatic variation with time, and it is this assumption that serves as the basis for using ancient vegetation to document long-term trends in climate. Short-term fossil sequences should provide the opportunity to capture vignettes that reflect the prevailing paleoclimate. Integration of such data from a number of sources should provide an index of the nature of climatic cycling. It is not our intention here to debate the validity of the "climax" concept, but rather to suggest that there are a number of factors related to these Neogene assemblages that greatly complicate this simplistic approach to documenting long-term climatic parameters. These complications involve short-term vegetation dynamics, initiated by external disturbance.

Disturbance-driven vegetation dynamics

While most ecologists view climate as a primary determinant in controlling the nature of vegetation in a given region, much attention is now focused on the effect of periodic and aperiodic disturbance, particularly fire, in controlling the distribution of specific vegetation types. In effect, external disturbance may act to keep the vegetation of a region in an effective state of ecological disequilibrium. Such considerations are critical in the analysis of Neogene vegetation in the Pacific Northwest, given the fact that the short temporal "windows" represented by the floras are intimately associated with local volcanism. The most obvious disturbance effects will be summarized below.

Hydrologic succession. To the extent that local volcanism diverts drainage channels and impounds "new" ponds and lakes, the effect is to initiate hydrologic successions on sites that had previously supported forest or other types of vegetation. The fossil stumps at Florissant represent trees (largely *Sequoia*) that were apparently inundated and killed as a consequence of the formation of "Lake Florissant." New lake basins will begin a sequence of eutrophication and infilling, providing that containment features are not breached prematurely. This eutrophication may leave a distinctive palynological signature in its early stages in terms of phytoplankton. Later stages, in which submerged and emergent aquatics may become significant, can typically be recognized both palynologically and from distinctive megafossil suites. Late eutrophication in the Neogene of the Pacific Northwest region typically involves development of taxodiaceous swamps dominated by *Glyptostrobus* or *Taxodium*. Closely spaced palynological samples can document the history of such a swamp; one such example, from the Whiskey Creek section at Succor Creek, is noted by Cross and Taggart (1982). A peak in *Taxodium* megafossils at the top of the P-33 section at Clarkia (Smiley and Rember, 1985) is another example of such a successional swamp. Such swamps often result in the deposition of lignites and organic-rich mudstones. Lignites throughout the Succor Creek region, taxodiaceous stump fields at the top of the Rockville and Shortcut sections at Succor Creek (Cross and Taggart, 1982), and mats of *Glyptostrobus* material in the Devils Gate section at Succor Creek (Satchell, 1984) reflect the late stages of hydrologic succession.

While the ecological dynamics of such successions are inherently interesting, such vegetation types were not necessarily common in the absence of volcanism, and the successions can be terminated at any point if the lake/pond containment features are breached. Such premature drainage appears to have been common in the Succor Creek area (Cross and Taggart, 1982). The abundance of either microfossil or megafossil material derived from such successions can be sufficient to cause a major distortion of relative-frequency fossil data. Since such seral stages are obviously unrelated to "climax" communities, such material must be factored out of any analysis that attempts to reconstruct the climax mosaic.

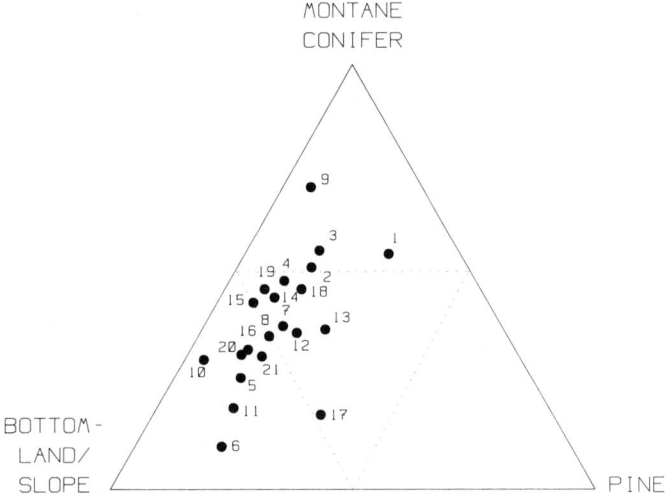

Figure 3. Ternary plot showing the range of variation in the relative representation of the bottomland/slope (broadleaved), montane conifer, and pine pollen elements in 21 samples from the 100-m Whiskey Creek section at Succor Creek.

Although hydrologic successions are relatively easily recognized, hydrologic processes associated with volcanic activity may have more subtle impacts on fossil assemblages. Figure 3 is a ternary diagram of 21 samples from the 100-m Whiskey Creek section in the Succor Creek area. The three composite pollen elements that are compared in this diagram include the bottomland/slope, montane conifer, and pine. The montane conifer element includes spruce (*Picea*), fir (*Abies*), and hemlock (*Tsuga*), with larch (*Larix*) and Douglas fir (*Pseudotsuga*) as rare components. The pine element includes all *Pinus* pollen types. Pine is a rare component in regional megafloras and is included as a single element in our pollen studies due to the wide ecological range of modern pines and the potential for long-range dispersal and consequent over-representation of pine pollen. The bottomland/slope element includes pollen of broadleaved trees and shrubs.

The Whiskey Creek pollen spectra show none of the characteristics attributable to direct volcanic disturbance but rather are marked by extreme variability in the relative importance of broadleaved and montane conifer pollen types, with dominance by the montane conifer element in almost half of the samples. Despite the rarity of conifer remains in the Succor Creek megafossil record, these pollen spectra suggest the proximal development of extensive conifer-dominated forests. The pervasive role of montane conifers in Succor Creek pollen spectra suggests that the prevailing paleoclimate was cool but highly equable, permitting the close juxtaposition of broadleaved taxa—some of which were frost-sensitive (*Oreopanax, Cedrela, Persea*)—in the lowlands, with conifer forests developed on local hillsides and uplands.

When the pollen spectra are plotted in stratigraphic sequence (Fig. 4), a cyclic sequence of broadleaved/conifer-dominated intervals is apparent. Early Whiskey Creek time appears to be particularly rich in terms of montane conifer forests; the amplitude of the cycles decreases in later Whiskey Creek time. We have previously suggested (Cross and Taggart, 1982) that such "cycling" may be related to small-scale thermal cycling, originally postulated by Milankovitch (1930) as a consequence of earth orbit mechanics and now considered to be a major causative factor in the climatic instability of the Pleistocene (Imbrie and Imbrie, 1979).

Although such cycling appears to be clearly reflected in the record, the extreme amplitude of the montane conifer peak at the base of the section may be related to a secondary effect of local volcanic activity. The base of the 100-m section consists of approximately 30 m of a relatively coarse, water-laid volcanic sandstone. The basal sample in this sequence is derived from a siltstone lens near the base of the measured section, the second sample is derived from a carbonate-rich layer near the middle of the sandstone unit, while the third sample is from a shale at the upper contact with the basal sandstone. All three samples are essentially identical in composition and very rich in montane conifer pollen. The similarity in the three samples suggested to us that the basal sandstone was deposited relatively quickly (Cross and Taggart, 1982); it is also possible that the magnitude of this conifer peak is an indirect consequence of volcanism.

The rapid deposition of 30 m of volcaniclastic sand must have inundated and destroyed broadleaved-dominated communities along the floor of the paleovalley. The destruction of these forests would result in a reduction in the relative production of broadleaved pollen, while conifer-dominated forests on adjacent slopes were largely unaffected, resulting in a major peak in montane conifer pollen in these basal spectra. Subsequent revegetation of the valley floor should have resulted in a gradual decline in the relative importance of montane conifer pollen. The subsequent peak in broadleaved pollen is the result of a major increase in alder (*Alnus*) pollen. Alders, which have the capability to fix nitrogen, often occur in nearly pure stands early in successional sequences following a variety of disturbances. It is found on disturbed riparian sites, is an early invader following glacial retreats in Alaska (Shelford, 1963), dominates alder scrub communities on volcanic ash in Japan (Tagawa, 1964), and dominates the early stages of shrub and tree establishment in many areas following fire and logging (Shelford, 1963; Harlow and Harrar, 1969). It is probable that the barren "sand flats" of the paleovalley must have developed such thickets following cessation of sand deposition. The secondary peak in montane conifer pollen levels probably represents another reduction in broadleaved pollen input, caused by senescence of the alder stands and their overtopping by reproductively immature hardwood forest. The following rise in broadleaved pollen is dominated by elm (*Ulmus*) pollen with peaks in hickory (*Carya*) and sweetgum (*Liquidambar*). This successional-forest mosaic in the valley bottom has a mixed-mesophytic aspect and is quite similar to the pattern of mixed-mesophytic–dominated spectra following possible fire events in the lower Devils Gate section in the northern Succor Creek area (Satchell, 1984). Spectra from the upper half of the section probably reflect ecologically mature forests, for the broadleaved element is quite diverse and co-dominated by elm and oaks. Such hydrologic disturbance is probably of limited extent under normal circumstances in upland/inland sites, but major stream inputs of massive quantities of ash may disrupt larger areas of forest, even in the absence of direct volcanic disturbance.

Direct volcanic disturbance. Direct blast effects, pyroclastic flows, lahars, and heavy ash falls all have the potential to destroy areas of preexisting forest and to initiate seral recovery leading back to relatively stable forest vegetation. We have devoted considerable attention to the consequence of direct volcanic disturbance at Succor Creek (Taggart and Cross, 1974, 1980; Cross and Taggart, 1982). Figure 5 shows a ternary diagram plotting 21 pollen/spore samples from the 30-m Rockville section at Succor Creek. The initial 13 samples (1 through 13) document the pre-disturbance vegetation, involving elevationally zoned, broadleaved, and montane conifer forest, essentially identical to the pattern observed in the Whiskey Creek section. The disturbance event occurs between sample 13 and 14, and sample 14 is dominated by pollen of herbaceous dicots (Compositae, Malvaceae, Chenopodiaceae, and Amaranthaceae) and grasses. The immediate post-disturbance sere is dominated by herbaceous plants that colonized the ash-covered landscape following destruction of the preexisting forests. In the Succor Creek area, this

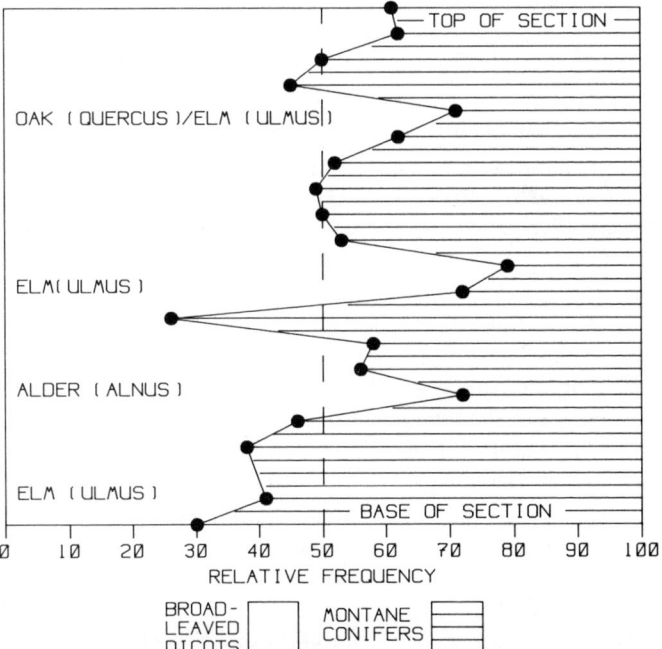

Figure 4. Variation in the representation of bottomland/slope (broadleaved) and montane conifer pollen elements in samples from the Whiskey Creek section at Succor Creek. The dominant taxa in the bottomland/slope element at various levels in the section are noted. Spruce (*Picea*) dominates the montane conifer element in all samples. The sample data are plotted sequentially with a constant interval between samples that does not reflect the actual stratigraphic intervals between samples.

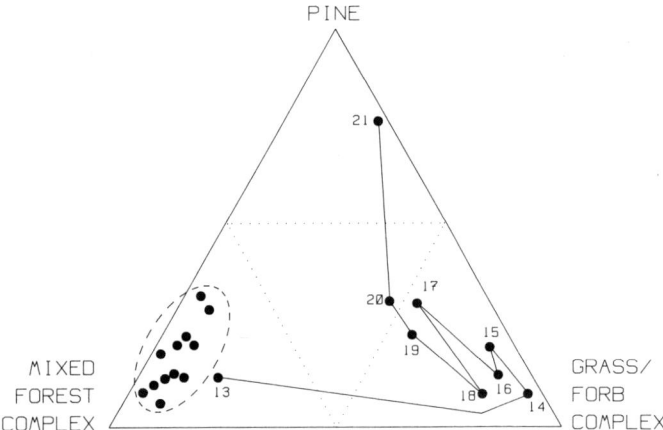

Figure 5. Ternary plot showing the variation in the relative frequency of the mixed forest (bottomland/slope and montane conifers), pine, and grass/forb pollen elements in samples from the 30-m Rockville section at Succor Creek. The samples are numbered sequentially (beginning with sample 13) from the base of the measured section. The initial 12 samples, characterizing the pre-disturbance forest vegetation, are contained within the ellipse on the left side of the diagram and are not numbered.

herbaceous seral stage is typically followed by what we have previously interpreted (Cross and Taggart, 1982) as a pine parkland. This interval is characterized by a pollen signature with elevated levels of pine pollen (notably in excess of the 5 to 10 percent pine pollen that represents normal background). These pines produced pollen types that can be differentiated from the pine pollen rain characteristic of the pre-disturbance forest. The transition to such a pine stage begins with sample 15, but between 15 and 16 there may be a second disturbance event that "resets" the sere to the "pure" grass/forb stage. Between 16 and 17, recovery begins again, but there is another seral setback between samples 17 and 18, probably reflecting another eruptive event. Local volcanic activity apparently subsides following sample 18, for recovery proceeds through samples 19, 20, and 21. The latter sample, the last productive sample from the upper Rockville section, represents a fully developed pine parkland. We do not consider this to be a closed-canopy pine forest stage; rather, we visualize it as consisting of scattered pines on a landscape otherwise clothed with grasses, annual and perennial "weedy" dicots, and woody shrubs. This stage may have similarities to open pine-scrub communities in the southeastern United States (Shelford, 1963, p. 73), which are associated with sandy soils and are successional to climax broadleaved forests in the process of soil stabilization and enrichment. Along the eastern shore of southern Lake Michigan, scrub pine forest is a characteristic stage in dune and sand-flat succession, prior to the establishment of more diverse broadleaved forests later in the successional sequence (Shelford, 1963, p. 53).

The post-parkland stage is not recorded at Rockville, but based on recovery data from the Type section to the north, the later recovery stages leading back to minimal-disturbance forest involve intermediate forests of mixed-mesophytic aspect in which *Carya, Liquidambar,* and *Ulmus* played a significant role (Taggart and Cross, 1980; Cross and Taggart, 1982).

Such volcanically induced successions are significant for several reasons. The first concerns time. The bulk of the time represented by the Rockville sediments documents seral recovery vegetation and not the pre-disturbance forest vegetation. The same is true for the Shortcut, Type, and Devils Gate sections at Succor Creek. In fact, of the five major sections we have studied in the Succor Creek region, only the Whiskey Creek section has no record of a major direct volcanic disturbance event. Based on this trend, it is reasonable to assume that the majority of the exposed volcaniclastic sediments in the Succor Creek area represent successional as opposed to "climax" vegetation.

Obviously one cannot assume that all volcaniclastic sequences will show a similar pattern, but it is equally obvious that it is not reasonable to make *a priori* assumptions regarding "climax" status for the source vegetation. The recognition of seral communities on the basis of fossil pollen and spores is relatively easy. Dominance by grasses or herbaceous pollen types and/or dominance of spectra by pine pollen appear to be significant recognition factors, coupled with low diversity of normal broad-leaved woody taxa. If stratigraphically controlled spectra are available, trends in the pollen data consistent with recovery and/or disturbance can serve to confirm the successional "diagnosis." It is important to emphasize that such determinations must be based on a considerable body of pollen data from the region in question. Conclusions derived from a small number of samples where background data are lacking are tentative at best.

Determination of successional status on the basis of megafossil data is more difficult, in part because enough productive levels through a specific section are more difficult to obtain. In the Succor Creek area, leaf-bearing horizons that are highly productive in number of specimens but comparatively low in diversity appear to be successional in nature.

Another disturbance signature that has been noted repeatedly in the Succor Creek area involves mats of conifer needles (primarily *Pseudotsuga* and *Picea*) in white airfall ashes in the Rockville and Shortcut sections and in multiple layers in the Coal Mine Basin at the southern end of the Succor Creek area. Pollen data suggest the presence of proximal conifer-dominated forests that were particularly well developed in the southern part of the Succor Creek area. Although montane conifer pollen can make up more than 50 percent of the pollen rain in such cases, the megafossil record there of such conifers is limited to a small number of seeds, a major factor in the omission of such conifers from previous vegetation reconstructions.

In marked contrast to these "normal" deposits, the conifer needle mats in these airfall ash beds were probably the result of facilitated needle fall, caused either by immediate blast effects or by mass needle fall following the nearly simultaneous death of a large number of conifers by outgassing or heat. In either case, the result was a large number of conifer needles flushed or blown into

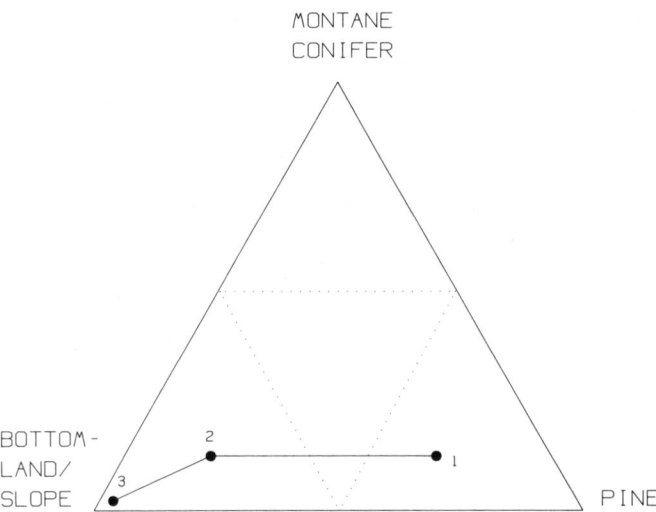

Figure 6. Ternary plot showing the variation in the bottomland/slope, montane conifer, and pine pollen elements in three samples from a 5-m section in the Stinking Water area. The samples are numbered sequentially from the base of the measured section.

the basin in a short period of time, resulting in distinctive mats of conifer debris, usually in concert with charcoal fragments. Such deposits almost certainly do not represent an increase in the importance of conifers in the local vegetation but rather are derived as a consequence of volcanic activity. They reflect a normal component of the vegetation of the watershed, but a component that is not normally well represented in the megaflora.

Our examination of pollen records from other floras in the region suggests that volcanic disturbance events similar to those at Succor Creek may be rather common in the Neogene of the Pacific Northwest. A short section from a Stinking Water locality is shown in ternary form in Figure 6. The sequence begins (sample 1) with high levels of pine pollen. Pine pollen declines steadily, and the sequence ends with a sample that is dominated exclusively by broadleaved pollen types. The most likely interpretation of this successional trajectory is a post-volcanic disturbance recovery where the first sample was derived from the pine-parkland stage. The minimal-disturbance "climax" at Stinking Water was probably a broadleaved forest in which montane conifers and pine were insignificant. Unfortunately, the extensive leaf collections from the various Stinking Water localities (Chaney and Axelrod, 1959) are not stratigraphically controlled, and new collections would be required to derive corresponding megafossil data. It is apparent from Figure 2 that the two major Stinking Water megafossil localities differ in the quantitative character of their plant assemblages, but the assessment of any ecological significance must await further study.

The Trapper Creek flora of southern Idaho is particularly notable among the Neogene floras of the Pacific Northwest by virtue of the fact that montane conifer remains make up approximately 25 percent of the several thousand specimens tabulated by Axelrod (1964) in his initial study of the flora. The Trapper Creek flora is quite diverse (66 taxa) and has been interpreted as an upland assemblage characterized by oak thickets (*Quercus simulata*) along the lake shore, intergrading with a complex mosaic of broadleaved-deciduous–dominated communities, and montane-conifer–dominated communities proximal to the basin. This interpretation is based on a collection of over 3,000 specimens composited from approximately 15 m of volcanic ash in an amphitheater-like exposure, created by terracing as part of a watershed improvement project. Another 30 m of section, stratigraphically below the amphitheater section, is exposed to the west, but these rocks, consisting of a monotonous series of airfall and water-laid ash deposits, are not particularly rich in megafossils and were not considered in Axelrod's study. The stratigraphic palynology of the composite 45-m section documents major patterns of vegetation dynamics that invalidate previous paleoecological reconstructions of this assemblage.

Plots of a number of these samples are included in Figures 7 through 9. The range of variation in a total of 29 productive samples from the 30-m basal ash unit of the West section is shown in Figure 7. These samples record a series of six major oscillations between low-diversity broadleaved-dominated and pine-dominated spectra (Fig. 8). It appears that early Trapper Creek time was characterized by at least five major episodes of volcanically induced forest disturbance. Montane conifers were apparently virtually absent from the area during this interval. Late in Trapper Creek time, in the 15-m amphitheater unit, disturbance apparently subsided sufficiently to permit a significant degree of uninterrupted forest recovery, as shown in Figure 9. Samples 1 through 7 (from the base upward) were recovered from the western half of the amphitheater exposure and show a distinctive successional trajectory. The first sample in this se-

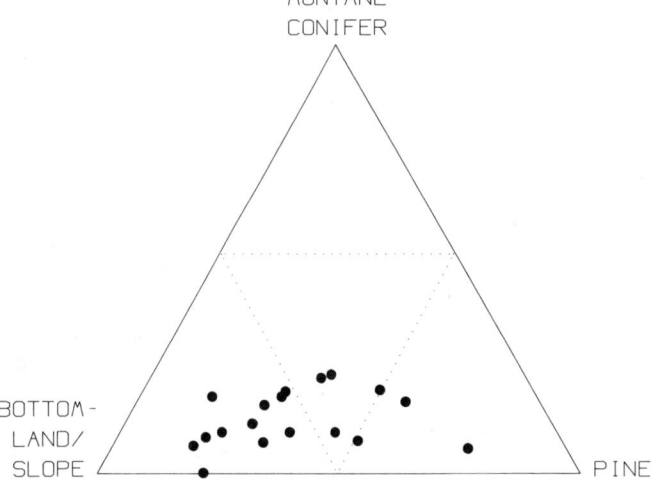

Figure 7. Ternary plot showing the variation in the bottomland/slope, montane conifer, and pine pollen elements in samples from the 30-m West section at Trapper Creek. These samples are plotted sequentially in Figure 8.

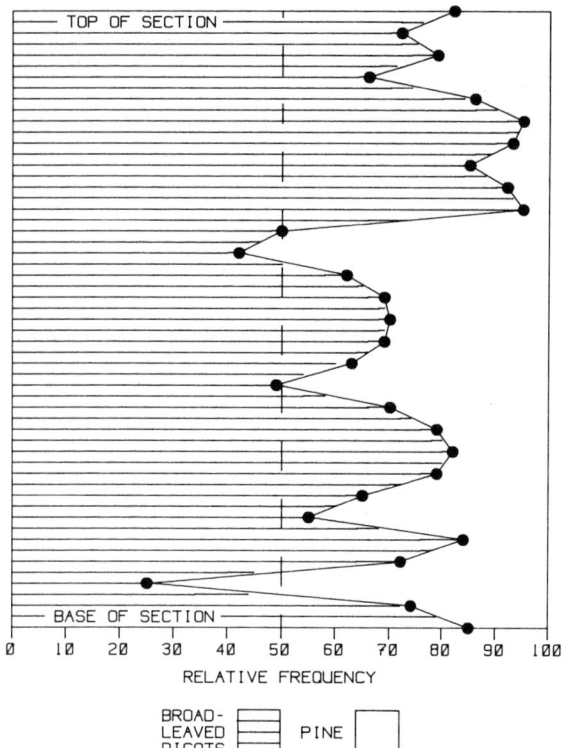

Figure 8. Variation in the relative frequency of broadleaved dicot and pine pollen elements through the 30-m West section at Trapper Creek. The samples are plotted sequentially with a constant vertical interval between samples that is not equivalent to the true stratigraphic interval between samples.

quence (1) is exclusively dominated by broadleaved taxa (*Ulmus, Quercus, Acer, Salix, Betula, Alnus,* and *Populus*) while samples 2 through 4 show progressive enrichment of the pollen spectra by montane conifer and pine pollen types. Samples 4 through 6 suggest that the source vegetation had reached an equilibrium characterized by a mosaic of broadleaved- and conifer-dominated communities. This reconstruction would be in general accord with Axelrod's analysis, but such a determination would be premature. Sample A was collected in the leaf-bearing ash at the east end of the amphitheater section and is quite different from any of the other spectra from the western half of the exposure. The spectrum of sample A is co-dominated by montane conifers and pine to the virtual exclusion of broadleaved pollen types. Field observations had suggested the possibility that the amphitheater beds were dipping to the east, so a section was sampled (B-E) about 200 m to the east to document the final stages of leaf-bearing ash accumulation as recorded at the Axelrod locality. The basal sample in this sequence (B) has a spectrum that is a balanced mixture of broadleaved, montane conifer, and pine pollen types, while the succeeding samples (C, D) are co-dominated by montane conifers and pine and are virtually identical to sample A. Sample E shows a return to a mixed spectrum.

Thus, it appears that the climax-forest mosaic at Trapper Creek consisted largely of montane conifers and some pine. This "climax," however, is only reached at one level—near the top of the east amphitheater section—based on the fossil pollen spectrum. This forest mosaic had been completely destroyed as a consequence of local volcanism in early Trapper Creek time, and repeated eruptions prevented any significant recovery. By early amphitheater-section time, activity apparently had subsided to some extent, and successional recovery, beginning with broadleaved-dominated forest communities, could begin. The successional trajectory toward the conifer-dominated "climax" was interrupted in a less complete fashion at least once. Whether this was due to proximal volcanism or fire is difficult to say. Fire is a distinct possibility given Satchell's (1984) observations of possible fire cycles in the lower Devils Gate section at Succor Creek, where mixed-pollen assemblages (mixed-mesophytic taxa, oaks (presumably evergreen) and conifers were repeatedly "reset" to mixed-mesophytic–dominated assemblages with gradual recoveries to the initial mixed assemblages.

Thus, the bulk of the amphitheater section, not to mention the earlier ash deposits, records succession in response to varying degrees of disturbance. Unfortunately, the original leaf collections were not documented as to individual stratigraphic occurrence in Axelrod's collections. The collections represent a mélange of seral stages, and while they are satisfactory for documenting the Trapper Creek floristic inventory, they are completely unsuited for ecological synthesis. It is interesting that Axelrod does recall (oral communication, 1987) that the bulk of the conifer material came from near the top of the leaf-bearing ash at the east end of the exposure, precisely where the pollen data suggest that conifer-rich layers should occur, i.e., near the end of the time of stratigraphic accumulation at the amphitheater site.

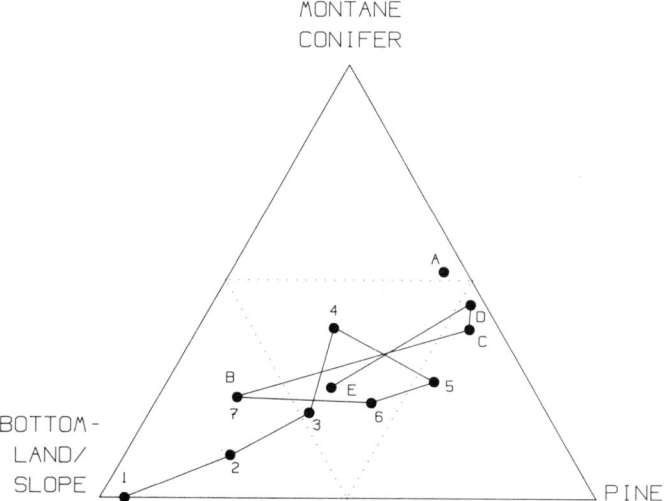

Figure 9. Ternary plot showing the variation in bottomland/slope, montane conifer, and pine pollen elements in samples from the Amphitheater exposure at Trapper Creek. Samples 1 through 7 were recovered from a 15-m sequence near the west end of the exposed outcrop and are numbered sequentially from the base of the measured unit: A is an isolated sample from the eastern end of the exposure; samples B through D were obtained from a 5-m measured sequence at the eastern end of the outcrop and are lettered sequentially from the base of the measured unit.

Volcanism and the faunal record. An interesting perspective is evident when the model of volcanically driven succession is applied to the problem of the large-mammal faunas associated with Neogene plant communities. There is no doubt that the Neogene "climax" vegetation throughout the Pacific Northwest involved a diverse array of forests. By any reckoning, the dominant large-mammal components of such communities should involve browsing herbivores. Shotwell (1968) defines two major faunal associations for the Miocene of the Great Basin. One of these, the *Hypohippus-Ticholeptus* association, is dominated by browsing herbivores but it is rare. The second, the *Merychippus-Dromomeryx* association, is common but is dominated by grazing herbivores. In a forested setting, browsing faunas should be far more common than grazing ones, but this would not appear to have been the case based on the available faunal record. Tens of thousands of specimens of leaves and other parts from forest trees and shrubs are compelling evidence of forest vegetation but mammalian dentition and trends toward cursorial habit are equally compelling evidence for a grazing mode of existence on the part of the large-mammal herbivores. This basic incongruity in the plant and mammal records (Taggart and others, 1982) has received remarkably little attention from paleobotanists, other than speculation regarding the possible existence of forb/grass-dominated habitats at some unspecified location on the paleolandscape. Given the impact of volcanically induced successions, however, the interpretive problems introduced by the grazing mammalian faunas on a forested landscape can be disposed of rather easily.

The faunas, like the floras, are preserved in volcaniclastic sediments, but it is important to recognize that the faunas do not typically occur at the same stratigraphic horizons as the floras. There is, for example, a small grazing-dominated fauna known from the Type section at Succor Creek. The bones are derived from a coarse fluvial unit near the middle of the section, and there are no co-occurring leaf fossils. Productive pollen samples above and below the bone beds (Cross and Taggart, 1982) indicate that the grazing mammals occurred during the grass/forb stage of a successional recovery following destruction of forests in the Type section area. In effect, the aspect of the vegetation of the Type section area during the period when the bones of the grazing assemblage were deposited was an open parkland, essentially identical to the post-disturbance recovery previously discussed for the Rockville section.

The pre-disturbance forest undoubtedly supported a browsing fauna, with the animals scattered throughout the forest. Minor eruptions that might have caused up to several centimeters of ash to accumulate in these forests would have minimal impact on such a fauna since the trees would not be killed and adequate browse would be available. Eruptions sufficiently severe to destroy the forest would also kill the majority of the large mammals. A few might survive to cluster along streams and around lakes where they would ultimately starve and perhaps be buried, but the number of such burials would be small compared with the animals killed in the initial eruptive event.

Following the initial eruption, the ash-covered landscape would be colonized by grasses and forbs and later by pine parklands. It was these landscapes that supported the grazing fauna. Such grazing associations were normally rare and probably followed successional patches of varying size and temporal duration. Major eruptions created seral patches, often of great extent, offering ideal habitats for grazing exploitation. Any such major eruptive episode is likely to be followed by secondary eruptions, a phenomenon consistent with the record at both Rockville at Succor Creek and the lower part of the Trapper Creek section. A heavy ash fall from such a secondary eruption would not necessarily kill the animals but would bury the grasses and forbs in a blanket of abrasive ash, condemning the grazers to death by starvation. It is highly likely that the stressed animal populations would gather around waterholes and along streams, probably in large numbers similar to those seen when drought strikes the modern savanna in Africa. Thus, large numbers of animals would die in proximity to depositional basins choked with reworked ash from the latest eruption. Such sites would have high potential for the preservation of large death assemblages. Thus, the nature of the volcanic eruptions and the ecologic consequences of such eruptive activity result in a situation where browsing faunas are rare while grazing assemblages are relatively well represented. This disparate pattern of representation will occur despite the fact that browsers are infinitely more common, in real terms, than grazers. Forest vegetation and an associated browsing fauna may dominate a landscape for millions of years, but in general, that time interval will not have a corresponding fossil record. In absolute terms, volcanic activity at any specific site is a rare occurrence, but such activity has the potential to create a fossil record that will preserve stages of vegetation development and associated faunas that are decidedly atypical. Such atypical successional communities will be differentially represented at the expense of more typical or "normal" community arrays. Such sampling is far from ideal for vegetation reconstruction, but it is an inescapable reality that must be recognized and compensated for in any paleoecological analysis.

SUMMARY

The volcaniclastic floras of the Neogene of the Pacific Northwest provide a record of vegetation that would probably not have been preserved under conditions of normal clastic sedimentation. While these floras represent a unique opportunity to study inland/upland vegetation throughout the region, their close spatial and temporal association with volcanic centers creates a major potential for disturbance-related vegetation dynamics, which complicates their interpretation. Modern approaches to the study of fossil floras emphasize the role of taphonomic factors in influencing the relation between source vegetation and the fossil assemblage, and such interpretive methodologies are entirely appropriate. Unlike extant vegetation studies, however, any paleoecological investigation involves time variables and the possibility of ecologically induced changes in source-vegetation composition and distribution, which have no relation to either evolution or long-term climatic change.

Evaluation of the impact of short-term ecological dynamics on a fossil assemblage requires close stratigraphic control, even in the context of a limited outcrop, regardless of the fossil material under study. Integration of stratigraphically controlled megafossil and pollen and spore data in conjunction with sedimentological data and taphonomic considerations would seem to provide the most comprehensive basis for reconstructing the source-vegetation mosaic and its dynamic attributes during the time interval represented by a specific exposure. In addition to providing fundamental information on the structure and dynamics of ancient plant communities, such information can serve to define the nature of minimal-disturbance communities on a specific site, providing a more reliable basis for the synthesis of long-term trends in vegetation distribution and paleoclimatic change.

REFERENCES CITED

Axelrod, D. I., 1964, The Miocene Trapper Creek flora of southern Idaho: University of California Publication in the Geological Sciences, v. 51, 180 p.
—— , 1966, A method for determining the altitudes of Tertiary floras: Paleobotanist, v. 14, p. 144–171.
—— , 1968, Tertiary floras and topographic history of the Snake River Basin, Idaho; Geological Society of America Bulletin, v. 79, p. 713–734.
—— , 1984, Biogeography of oaks in the Arcto–Tertiary province: Annals of the Missouri Botanical Garden, v. 70, p. 629–657.
Axelrod, D. I., and Bailey, H. P., 1969, Paleotemperature analysis of Tertiary floras: Palaeogeography, Palaeoclimatology and Palaeoecology, v. 6, p. 163–195.
Chaney, R. W., 1925, The Mascall flora; Its distribution and climatic relation: Carnegie Institute of Washington Publication 349, p. 23–48.
—— , 1948, The bearing of the living Metasequoia on the problems of Tertiary paleobotany: Proceedings of the National Academy of Science, v. 34, p. 503–515.
—— , 1951, A revision of the fossil Sequoia and Taxodium in western North America based on the recent discovery of Metasequoia: Transactions of the American Philosophical Society, v. 40, p. 171–263.
Chaney, R. W., and Axelrod, D. I., 1959, Miocene floras of the Columbia Plateau: Carnegie Institute of Washington Publication 617, 237 p.
Cross, A. T., and Taggart, R. E., 1982, Causes of short-term sequential changes in fossil plant assemblages; Some considerations based on a Miocene flora of the northwest United States: Annals of the Missouri Botanical Garden, v. 69, p. 679–734.
Graham, A., 1965, The Sucker Creek and Trout Creek Miocene floras of southeastern Oregon: Kent, Ohio, Kent State University Bulletin of Research, Series IX, v. 53, 147 p.
Harlow, W. M., and Harrar, E. S., 1969, Textbook of Dendrology, 5th ed.: New York, McGraw-Hill, 512 p.
Imbrie, J., and Imbrie, K. P., 1979, Ice ages: New Jersey, Enslow Publishers, 224 p.
MacGinitie, H. D., 1941, The middle Eocene floras from the central Sierra Nevada: Carnegie Institute of Washington Publication 534, 178 p.
McLeroy, C. A., and Anderson, R. Y., 1966, Laminations of the Oligocene Florissant lake deposits, Colorado: Geological Society of America Bulletin, v. 77, p. 605–618.
Milankovitch, M., 1930, Mathematische Klimalehre und der astronomische Theorie der Klimaschwankungen, in Koppen, W., and Geiger, R., eds., Handbuch der Klimatologie, I(A): Berlin, Gebruder Borntraeger, p. 1–176.
Satchell, L. S., 1984, Patterns of disturbance and vegetation change in the Miocene Succor Creek flora (Oregon–Idaho) [Ph.D. thesis]: East Lansing, Michigan State University, 153 p.
Shelford, V. E., 1963, The ecology of North America: Urbana, University of Illinois Press, 610 p.
Shotwell, J. A., 1968, Miocene mammalian faunas of southeastern Oregon: Bulletin of the University of Oregon Museum of Natural History, v. 14, p. 1–67.

Smiley, C. J., and Rember, W. C., 1985, Composition of the Miocene Clarkia flora, in Smiley, C. J., ed., Late Cenozoic history of the Pacific Northwest; Interdisciplinary studies on the Clarkia fossil beds of northern Idaho: San Francisco, California Pacific Division American Association for the Advancement of Science, p. 95–112.
Spicer, R. A., 1981, The sorting and deposition of allochthonous plant material in a modern environment at Silwood Lake, Silwood Park, Berkshire, England: U.S. Geological Survey Professional Paper 1143, 77 p.
Spicer, R. A., and Wolfe, J. A., 1987, Plant taphonomy of late Holocene deposits in Trinity (Clair Engle) Lake, northern California: Paleobiology, v. 13, p. 227–245.
Tagawa, H., 1964, A study of the volcanic vegetation in Sakurajima, southwest Japan: Memoirs of the Faculty of Science, Kyushu University, Series E, Biology, v. 3, p. 166–288.
Taggart, R. E., 1971, Palynology and paleoecology of the Miocene Sucker Creek flora from the Oregon–Idaho boundary [Ph.D. thesis]: East Lansing, Michigan State University, 196 p.
—— , 1973, Additions to the Miocene Sucker Creek flora of Oregon and Idaho: American Journal of Botany, v. 60, p. 923–928.
—— , 1987, The effect of vegetation heterogeneity on short stratigraphic sequences, in DiMichele, W. A., and Wing, S. L., eds., Methods and applications of plant paleoecology; Notes for a short course: Paleobotanical Section, Botanical Society of America, p. 184–211.
Taggart, R. E., and Cross, A. T., 1974, History of vegetation and paleoecology of upper Miocene Sucker Creek beds of eastern Oregon: Birbal Sahni Institute (Lucknow) Special Publication 3, p. 125–132.
—— , 1980, Vegetation change in the Miocene Sucker Creek flora of Oregon and Idaho; A case study in paleosuccession, in Dilcher, D. L., and Taylor, T. N., eds., Biostratigraphy of fossil plants: Stroudsburg, Pennsylvania, Dowden, Hutchinson, and Ross, Inc., p. 185–210.
Taggart, R. E., Cross, A. T., and Satchell, L., 1982, Effects of periodic volcanism on Miocene vegetation distribution in eastern Oregon and western Idaho; Proceedings of the Third North American Paleontology Convention, Montreal: Toronto, Business and Economic Service Ltd., v. 2, p. 535–540.
Wolfe, J. A., 1971, Tertiary climatic fluctuations and methods of analysis of Tertiary floras: Palaeogeography, Palaeoclimatology, Palaeoecology, v. 9, p. 27–57.
—— , 1978, A paleobotanical interpretation of Tertiary climates in the northern hemisphere: American Scientist, v. 66, p. 694–703.
—— , 1979, Temperature parameters of humid to mesic forests of eastern Asia and relations to forests of other regions in the northern hemisphere and Australia: U.S. Geological Survey Professional Paper 1106, 37 p.

MANUSCRIPT ACCEPTED BY THE SOCIETY JUNE 21, 1989

Taphonomy and sedimentology of Arikaree (lower Miocene) fluvial, eolian, and lacustrine paleoenvironments, Nebraska and Wyoming; A paleobiota entombed in fine-grained volcaniclastic rocks

Robert M. Hunt, Jr.
Division of Vertebrate Paleontology, University of Nebraska, Lincoln, Nebraska 68588-0514

ABSTRACT

Nonmarine lower Miocene rocks widely exposed in nearly continuous outcrop over approximately 3100 km^2 (1,200 mi^2) of the Hartville Table in southeastern Wyoming and western Nebraska indicate a semiarid continental interior, with seasonal climate characterized by sandy ephemeral or intermittent braided streams, interchannel plains mantled by fine-grained volcaniclastic loess, and shallow ephemeral holomictic lakes. These paleoenvironments are recognized on the basis of distinctive sedimentologic, faunal, and taphonomic characteristics.

Stream sediments (10 percent or less of total outcrop) are primarily tuffaceous silty sandstones, deposited as reworked pyroclastic debris in wide shallow valleys. These valleys first filled with fluvial fine-grained volcaniclastics, but with the cessation of streamflow in the region, filling was completed by air-fall volcaniclastic loess that blanketed both valleys and interchannel reaches. Fluvial sediments within the valleys include much spatially dispersed mammal bone that had been scavenged and subaerially weathered prior to burial. Waterholes, situated in or adjacent to the valleys, filled with tuff and carbonate mud containing freshwater ostracods, pulmonate gastropods, diatoms, and charophyte algae. These tuffaceous waterhole muds intertongue with fluvial volcaniclastic sediments and are the locus of major mammalian bone beds, the best known preserved at Agate Fossil Beds National Monument. Bones of chalicothere, rhinoceros, and entelodont are common in waterhole bone beds and in fluvial sediments in the region.

Massive tuffaceous air-fall silty sandstones (87 percent of outcrop) punctuated by silcrete paleosols were deposited in the interchannel reaches; mammal remains are commonly represented by widely scattered, isolated bones and partial skeletons of young and aged ungulates, chiefly oreodonts and camels, indicative of attritional deaths over time. No bone beds occur.

Thin silicified carbonate mudstones (about 2 percent of outcrop) with ostracods, plant debris, and aquatic pulmonate gastropods (but without fish or other aquatic vertebrates) indicate shallow, holomictic, ephemeral lakes that filled with homogeneous micrite mud. These lakes were isolated sheet-like bodies of water unassociated with stream sediments. Following desiccation, lacustrine sediments were commonly overprinted by pedogenic features.

Eolian transport of fine pyroclastic detritus into the North American midcontinent was essential to preservation of these sedimentary environments and their rich fossil

Hunt, R. M., Jr., 1990, Taphonomy and sedimentology of Arikaree (lower Miocene) fluvial, eolian, and lacustrine paleoenvironments, Nebraska and Wyoming; A paleobiota entombed in fine-grained volcaniclastic rocks, *in* Lockley, M. G., and Rice, A., eds., Volcanism and fossil biotas: Boulder, Colorado, Geological Society of America, Special Paper 244.

record. In the Americas and in Africa during the Cenozoic, fine-grained volcaniclastic sediments blanketed large geographic areas within the continental interiors, preserving significant temporal intervals of the vertebrate fossil record. If volcanism had not occurred, these intervals would exist as major hiatuses in our knowledge of vertebrate, particularly mammalian, evolution. The important role of fine-grained volcaniclastics in preservation of mammalian faunas and their associated depositional environments in the Americas and in Africa during the Cenozoic deserves greater emphasis.

INTRODUCTION

Today a growing awareness of the importance of volcanism in the preservation of terrestrial biotas is developing from integrated study of both fossils and continental sediments worldwide. In Africa, initial studies by W. W. Bishop (1963, 1968) emphasized the importance of fine-grained subaerially deposited volcaniclastics, particularly primary air-fall carbonatite ash, to the preservation of terrestrial vertebrates. Bishop envisioned the burial of attritional skeletal debris, much of it scavenged and subaerially weathered on land surfaces, by periodic ash falls, which because of their chemistry, contributed to the development of alkaline soils conducive to bone preservation. More recent work in East Africa (Hay, 1986; Pickford, 1986) has confirmed Bishop's insight, suggesting that not only is volcanic ash important to the burial of bone in a wide range of depositional environments, but also that its chemistry promotes bone mineralization. In South America, the importance of fine-grained volcaniclastics to preservation of the terrestrial vertebrate fossil record has long been evident (Hatcher, 1903; Pascual and others, 1985), yet in-depth study of the relation between environments of deposition and fossil occurrence is in its infancy. In North America, the contribution of fine-grained volcaniclastics to the preservation of the rich continental record of Cenozoic fossil mammals was first appreciated early in this century (Matthew, 1915) but subsequently has received little attention. Especially noteworthy is the importance of volcaniclastics to the preservation of the North American Oligocene and early to mid-Miocene mammalian record, particularly in the midcontinent. Thus, my intent in this chapter is to resurrect an awareness of the importance of fine-grained volcaniclastics to the preservation of the North American Cenozoic mammal record; to demonstrate the wide range of depositional environments that are recorded by such deposits; and to present an example from the early Miocene of the North American midcontinent that indicates how lithofacies can preserve faunal distribution patterns, just as such patterns have been recently revealed in the African Miocene (Pickford, 1986).

Detailed facies analysis of volcaniclastic[1] sediments in the rock record has only recently begun to receive attention (Fisher and Schmincke, 1984; Hay, 1976; Lajoie, 1984; Pickford, 1986; Schmid, 1981; Smith, 1986; Suthren, 1985). Facies models developed to date are primarily concerned with proximal volcaniclastic sedimentation associated with convergent plate margins and rifts. The nature of lithofacies in thick volcaniclastic accumulations of loessic materials initially deposited as fine pyroclastic debris distal to the source terrain is largely untreated (Vicars and Breyer, 1981; Hunt, 1985).

Volcaniclastic loess is herein defined as regionally widespread, often thick, accumulations of uniformly fine-grained tuffaceous sediment, comprising chiefly silt and very fine to fine sand, with a significant percentage of volcanic glass shards and/or pyrogenic mineral crystals. The textural uniformity of volcaniclastic loess in combination with its high shard/pyrogenic crystal content indicates a pyroclastic origin for the bulk of the sediment. Volcaniclastic loess is often massive or structureless; the lack of stratification appears to result from airfall on level terrain with sufficient plant cover to act as a baffling mechanism preventing dune or sand sheet formation. Extensive bioturbation by burrowing animals and plant roots also is a probable factor in homogenizing the sediment, given the average sediment accumulation rates (from 2 to 3 cm/1,000 yr using geophysical methods, to 0.5 to 1 m/1,000 yr using estimates of soil formation; see Retallack, 1983, and calculations in this report) estimated for volcaniclastic loess bracketed by vitric tuffs in the North American midcontinent.

Where wind and/or water have converted primary airfall volcaniclastics into stratified deposits, the resulting sediments are termed, respectively, eolian and/or fluvial volcaniclastics. Such reworked pyroclastics are commonly associated with volcaniclastic loess in the central Great Plains. When stratified volcaniclastics occur in association with volcaniclastic loess, the stratified eolian and fluvial units contain fewer vitric shards, are better sorted, and show increased grain roundness, while retaining similar kinds and relative amounts of constituent grains, suggesting that the loess has undergone transport and sorting and, in fact, is the parent material from which the stratified volcaniclastics are derived.

Volcaniclastic loess deposits are significant in that their sheet-like geometry, developed over large geographic areas, and their considerable thickness, produced by gradual aggradation over long intervals of time, contribute to the preservation of important terrestrial biotas of vertebrate, invertebrate, and plant remains in the fossil record. Furthermore, long-term survival of these fossil biotas at the scale of geologic time may be enhanced through accumulation of such loessic deposits in the interior of continents or in other quiescent tectonic settings where preservation potential is higher than at destructive plate margins.

---

[1]The term *volcaniclastic* as used in this chapter follows the definition of Fisher and Schmincke (1984, p. 89): "all clastic volcanic materials formed by any process of fragmentation, dispersed by any kind of transporting agent, deposited in any environment or mixed in any significant portion with nonvolcanic fragments."

TERRESTRIAL VOLCANICLASTIC LOESS IN THE AMERICAS

During the Cenozoic, as the result of plate tectonic interactions at the western margins of the Americas, fine-grained volcaniclastics were extensively deposited in the continental interiors where they were instrumental in preservation of important land mammal faunas of Eocene, Oligocene, and early to mid-Miocene age in North America, and Paleocene through Miocene age in South America. Were it not for these volcaniclastic sequences, our knowledge of pattern and species diversity in New World land mammal evolution would be seriously limited during these intervals.

Thick, nonmarine, mammal-bearing volcaniclastic sequences, especially in the mid-Cenozoic, resulted from accumulation of enormous volumes of air-borne volcanic ash carried by westerly winds eastward to open lowland settings marginal to the uplifts. Deposition of fine-grained volcaniclastics built level land surfaces occupied by woodland, savanna woodland–parkland, and grassland environments. Reworking of gradually aggrading volcaniclastics by wind and water in these terrestrial environments, and mixing with fine-grained epiclastic materials from nearby uplifts, was commonplace.

In North America, fine-grained volcaniclastics make up the greater part of the White River (Sinclair and Granger, 1911; Wanless, 1922, 1923; Retallack, 1983; Emry and others, 1987) and Arikaree (Darton, 1899; Hunt, 1985; Swinehart and others, 1985) Groups of the North American midcontinent, as well as the John Day and Mascall Formations of Oregon (Merriam, 1901; Fisher and Rensberger, 1972). White River and Arikaree rocks in the midcontinent attain thicknesses of nearly 460 m (1,500 ft) (Darton, 1899) and include enormous amounts of rhyolitic ash; the John Day Formation comprises about 760 to 1220 m (2,500 to 4,000 ft) of silicic volcaniclastic rocks, predominantly air-fall rhyolitic and dacitic ash that weathered to montmorillonite during subaerial exposure (Hay, 1963). These sediments contain the major Oligocene and early Miocene reference faunas of land mammals used to calibrate the nonmarine rock record in North America (Tedford and others, 1987).

In South America, most of the Paleogene and early Neogene mammal-bearing rocks of Patagonia are fine-grained volcaniclastics (Pascual and others, 1985). The principal samples of South American fossil land mammals from the Paleogene through early Neogene were derived from Patagonian deposits (Marshall and others, 1983). The importance of volcaniclastic deposition to the early Cenozoic land mammal record of South America is now well known (Pascual and others, 1985; Patterson and Pascual, 1972, p. 251). Recent field studies by T. M. Bown (written communication, 1988) and his associates indicate that during Santacrucian time (early to mid-Miocene; Marshall, 1985), an interval distinguished by its abundant and well-preserved mammalian fossils, pyroclastic depositional centers with little associated fluvial activity bordered the Andean Precordillera on the east. Along the Atlantic coast, pyroclastics are less in evidence, and a fluvially dominated volcaniclastic sequence is present.

In North America, White River and Arikaree volcaniclastics accumulated on flat, gently east-sloping geomorphic surfaces that extended from the front ranges of the Rocky Mountains into the midcontinent, and also filled Rocky Mountain basins adjacent to uplift cores. Climate ranged from warm subtropical in early White River time to warm semiarid in later White River and Arikaree time. The John Day volcaniclastics were probably deposited under more humid conditions on a broad lowland bordered by adjacent highlands: "Topographically, the basin was a large and nearly featureless plain with interior drainage bordered by low-standing ash-mantled hills" (Fisher and Rensberger, 1972, p. 19). In South America, the Santa Cruz Formation, the most prolific producer of fossil mammals on the continent, includes as much as 730 m (2,400 ft) of tuffaceous sediments widespread in southern Patagonia (Russo and others, 1980). A warm humid climate is indicated, prior to the drying effect and development of savannas created by the later elevation of the Andes.

White River and Arikaree deposition has been dated at about 37 to 19 Ma (references in Emry and others, 1987; Tedford and others, 1987); John Day deposition extends from about 32 to ~18 Ma (Tedford and others, 1987); deposition of Patagonian volcaniclastics extends from at least 60 to 5 Ma (Pascual and others, 1985, p. 221–222; Bown, written communication, 1988).

The approach first adopted in the initial geological and paleontological studies of these terrestrial volcaniclastic sequences was descriptive and stratigraphic. Early pioneering investigations were in keeping with the style of field geology practiced in the late nineteenth and early twentieth century when these rock sequences were first extensively explored and their fossils discovered and described. Work by Hayden (1857; Meek and Hayden, 1861), Darton (1899), Hatcher (1893, 1902), Matthew (1901), Peterson (1907, 1909), and Wortman (1893) in the Arikaree and White River; Merriam (1901) and Calkins (1902) in the John Day; and Carlos Ameghino in the late 19th century in Patagonia (Simpson, 1984) established a basic stratigraphic framework and demonstrated that these beds were sources of abundant mammalian fossils. Fossils were first discovered in these volcaniclastics (White River, 1840s; Arikaree, 1850s; John Day, 1861; Santa Cruz, 1887) prior to the establishment of a firm lithostratigraphy. Thus, stratigraphic levels of the earliest collections were not always accurately determined because of the reconnaissance nature of the initial explorations.

In general, the first attempts at subdivision of the thick volcaniclastic sequences were not based on well-defined or detailed lithologic criteria. In the field, the volcaniclastic loess sequences displayed a vertical and lateral homogeneity that proved difficult for earlier workers to subdivide. Sharp bounding contacts of regional extent were rarely observed, and when seen, their significance was often misunderstood. Consequently, mammalian fossils frequently proved to be the most expedient tools for subdivision (Matthew, 1899a), and subunits initially were based on biostratigraphic zonation (e.g., Titanotherium beds, Oreodon beds, Diceratherium beds), often supplemented by the available lithologic criteria.

Despite the lack of resolution of much early work, particularly well-reasoned analyses of the depositional environments of fine-grained terrestrial volcaniclastics in western North America were published by Matthew (1899b, 1901), Merriam (1901), and Hatcher (1902).

Recognition of the significant volcaniclastic contribution to the widespread, richly fossiliferous, nonmarine Tertiary sequences of western North America was slow in developing (Sinclair, 1912; Johannsen, 1914; Matthew, 1915). This realization grew out of the collapse of the lacustrine theory of origin of the western North American Tertiary that had been propounded by King, Hayden, and Darton, among others, in the latter half of the 19th century. In now-classic studies published at the turn of the century, based on careful field observations, the paleontologists Matthew (1899b, 1901) and Hatcher (1902) argued on strong geological and paleontological evidence that the White River Oligocene beds of the Great Plains were the product of eolian and fluvial, not lacustrine, processes within a subhumid to semiarid continental interior. In fact, Matthew asserted that a large part of the White River Formation of Colorado accumulated as wind-blown loess, a view generally accepted today for at least the upper part of the formation (Emry and others, 1987). A few years earlier, following detailed field investigations, Gilbert (1896) and Haworth (1897) had attributed the major part of the Tertiary of eastern Colorado and western Kansas to fluvial depositional processes, thereby initiating abandonment of the lacustrine theory.

Almost simultaneously, field investigations in the western United States, prompted by the discovery of rich fossil vertebrate accumulations, led to pioneering petrographic studies (Calkins, 1902; Sinclair, 1906, 1909; Sinclair and Granger, 1911) that demonstrated the major volcaniclastic component of rocks filling a number of the Tertiary basins. Earlier, Darton (1899) noted the large amount of volcanic ash incorporated in Arikaree Group sediments of the Great Plains. Merriam (1901), based on Calkins' petrographic work (published in 1902), determined that the John Day Formation as well as sub- and superjacent Oligocene and Miocene rocks in Oregon were composed of tuffaceous sediments and ash beds. He observed evidence for both eolian and fluvial reworking of fine-grained tuffaceous sediments in the John Day basin; believed that direct air-fall deposition of unreworked tuffaceous units had occurred; and thought that a loessic origin for some John Day sediments was plausible.

By 1912, W. J. Sinclair of Princeton University could write, "The important part played by volcanic ash in [the lithology of western Tertiary formations] has only recently been realized. We now know that the Santa Cruz formation of Patagonia is composed entirely, or almost entirely, of ash, while of North American horizons, the Bridger, Washakie, John Day, Mascall, Rattlesnake, a large part of the White River, and probably others not investigated, are either in large part or entirely ash."

The full import of the volcanic contribution to the continental Tertiary of western North America was summarized a year later by W. D. Matthew (1915) in the Silliman Lectures at Yale in December 1913: "There is one element that is not today contributing to any appreciable extent to the formations of the Cordilleran regions . . . but during the Tertiary was almost the principal source of the terrestrial sediments. This is volcanic ash. . . . Volcanic ash is a principal source of a large part of the Tertiary formations of the plains." Matthew emphasized that, whereas a number of western interior basins in North America were filled with Tertiary epiclastic deposits derived from bordering uplifts, other basin fills were chiefly volcaniclastic. Significantly, his insight directly resulted from the first applications of petrographic analysis to fossiliferous, nonmarine, Tertiary basin fills in western North America by men such as Calkins, Darton, and Sinclair.

LOCATION AND AIMS OF STUDY

Recent work on Cenozoic rocks in the North American midcontinent (Diffendal, 1982; Gustavson and Winkler, 1988; Hunt, 1978, 1985; Swinehart and others, 1985; Tedford and others, 1985; Vicars and Breyer, 1981) has attempted to establish an objective lithostratigraphy using lithology, facies analysis, marker beds (especially paleosols and tuffs), and regionally significant horizons of erosional and depositional discontinuity, such as widespread unconformities and changes in style of sedimentation. Such criteria can be used to define lithostratigraphic units having genetic and temporal integrity in the geologic history of the region.

My purpose in this chapter is to suggest the potential for objective subdivision and lithofacies identification within thick, nonmarine, loessic volcaniclastic sequences, and to demonstrate the usefulness of combined lithologic and faunal data in reconstructing depositional environments of these fine-grained volcaniclastic rocks. In addition, the vertical succession of mammalian biofacies within these volcaniclastic sequences provides important evidence of the tempo and mode of evolution over the large geographic areas blanketed by such sediments. The term "facies," as used in this study, is defined as "the sum total of lithological and faunal characteristics of a given stratigraphic subunit" (Hallam, 1981). Once facies are defined, and their associations in outcrop identified, not only can depositional environments be interpreted but the organization of facies into more objectively defined lithostratigraphic units becomes possible.

My work has focused on the Arikaree Group (late Oligocene to early Miocene) of the North American midcontinent in its type area in the central Great Plains, because of its wide geographic distribution, good exposures, excellent preservation of fossil vertebrates in a variety of depositional environments, presence of datable tuffs, and predominance of fine-grained volcaniclastic rocks amenable to paleomagnetic sampling. Originally mapped and described by Darton (1899), and further subdivided on lithic and faunal criteria by Hatcher (1902), no recent effort had been made to objectively identify subdivisions of this 180 to 215-m-thick (600 to 700 ft) sequence of fine-grained volcaniclastic sediments. Currently available subsurface data are not adequate to permit detailed lithofacies analysis. However, the present distribution of Arikaree outcrops over the surface of a broad elevated tableland east of the Hartville uplift of southeastern Wyoming is well suited to the recognition of lithofacies within rocks of the upper Arikaree Group (early Miocene). This has led to the identification of a regionally extensive, upper Arikaree formation-rank unit, which is well exposed over much of the surface of the tableland (Hunt, 1978, 1981, 1985). Only a single stratotype among those designated by the early stratigraphers in this region corresponds to this formation: the "Upper Harrison" beds (Peterson, 1907, 1909), a poorly chosen name applied to these rocks following the realization that the name first used for them (Nebraska Beds, Hatcher, 1902) was invalid. I continue to employ this name, pending substitution of a more appropriate lithostratigraphic term, because it unambiguously identifies this formation using the term applied to it by Peterson on the basis of the earliest designated stratotype. Accordingly, in this report the "Upper Harrison" beds and the Harrison Formation constitute the two formations making up the upper Arikaree Group (Hunt, 1985). In the North American land mammal biochronology, the upper Arikaree Group is characterized by the Late Arikareean chronofauna of early Miocene age (Hunt, 1985; Tedford and others, 1987).

GEOGRAPHIC AND GEOLOGIC SETTING

In the type area in western Nebraska, southeastern Wyoming, and southwestern South Dakota, the White River and Arikaree Groups comprise about 460 to 760 m (1,500 to 2,500 ft) of directly superposed, fine-grained, volcaniclastic rocks. These rocks are the remnant of a great Cenozoic clastic wedge that extended eastward from the Rocky Mountains into the midcontinent during the Oligocene and Miocene, its lower part largely built up by uniformly fine-grained volcaniclastic sediments, and its surface veneered by a thin, texturally varied, predominantly fluvial epiclastic unit, the Ogallala Formation (for Ogallala references, see Diffendal, 1982, 1984; Goodwin and Diffendal, 1987). Plio-Pleistocene erosion (Darton, 1899; Trimble, 1980) incised the wedge, removing an enormous volume of sediment, so that eventually only two extensive but dissected tablelands remained in the central Great Plains (Fig. 1). The margins of these tablelands are vertical escarpments in which White River–Arikaree strata are well exposed, and lateral continuity can be demonstrated over considerable distances.

Arikaree rocks are best developed along the bounding escarpments and over the surface of the Hartville Table (Fig. 2) of southeastern Wyoming and western Nebraska. Here, dissection has exposed the uppermost lithostratigraphic unit of the Arikaree Group, the Upper Harrison beds of Peterson, over an area of approximately 3100 km^2 (1,200 mi^2) (Fig. 3). The lithostratigraphic identity of the formation can be demonstrated by identification of a particular lithofacies association bracketed between bounding contacts with sub- and superjacent units (Fig. 4). The Upper Harrison stratotype of Peterson (1909) is a particularly well-exposed section of about 125 feet (38 m) along the south escarpment of the Niobrara Canyon immediately east of the Wyoming-Nebraska border (the south rim of the canyon in T31N,R57W, Sioux County, Nebraska, can be established as Peterson's type area). Present in the stratotype area are all major lithofacies identified in the formation. The Upper Harrison beds are separated from subjacent Arikaree rocks by a sharp lower-bounding contact of regional extent, expressed either as an erosional disconformity or as a depositional hiatus associated with development of a major paleosol (Hunt, 1978, Fig. 3; 1985, Fig. 3). An erosional episode is indicated at many localities where the terminal Harrison paleosol horizon is cut out and Upper Harrison fluvial sediments deposited as fill. The upper contact with younger rocks of the Hemingford Group is exposed in the central part of the tableland along the valley of the Niobrara River in the vicinity of the Agate Fossil Beds National Monument, but is absent elsewhere on the tableland where these overlying rocks have been removed by late Cenozoic erosion.

Between these bounding contacts, Upper Harrison lithologies are predominantly tuffaceous silty sandstone, carbonate mudstone, very rare claystone and diatomite, intraformational conglomerate, and vitric tuff. The greater part of the formation is tuffaceous sandstone, characterized by fine grain size (chiefly silt and very fine sand), generally good sorting, loose grain packing, predominance of subangular to angular grains, presence of abundant vitric shards, friable texture, light brownish gray color (10YR 6.5/2), and minimal diagenetic alteration. Diagenesis is primarily limited to slight alteration of grain surfaces; solution of some chemically unstable grains; and emplacement of calcite, montmorillonite, and silica cements through ground-water migration and pedogenic processes (Stanley and Benson, 1979). No zeolites have been identified.

The absence of pronounced diagenetic modification of these sediments, never deeply buried and fortuitously deposited in a tectonically quiescent setting, coupled with early precipitation of void-filling cement, has been responsible for the excellent state of preservation of vertebrate fossils in these and other Arikaree rocks. Vitric sandstones of the Arikaree Group display only minor hydration and dissolution of volcanic glass shards; this appears to be possible only if contact with interstitial pore fluids was infrequent, and pore fluids were dilute waters of near-neutral pH (Stanley and Benson, 1979). Once buried in Arikaree sandstone, vertebrate bone remained largely uncompacted, retained its hydroxyapatite composition and microstructure (Rogers, 1924) and, in the void spaces created by the decay of organic material, was often impregnated with calcite or silica in an alkaline to pH-neutral subsurface environment soon after burial.

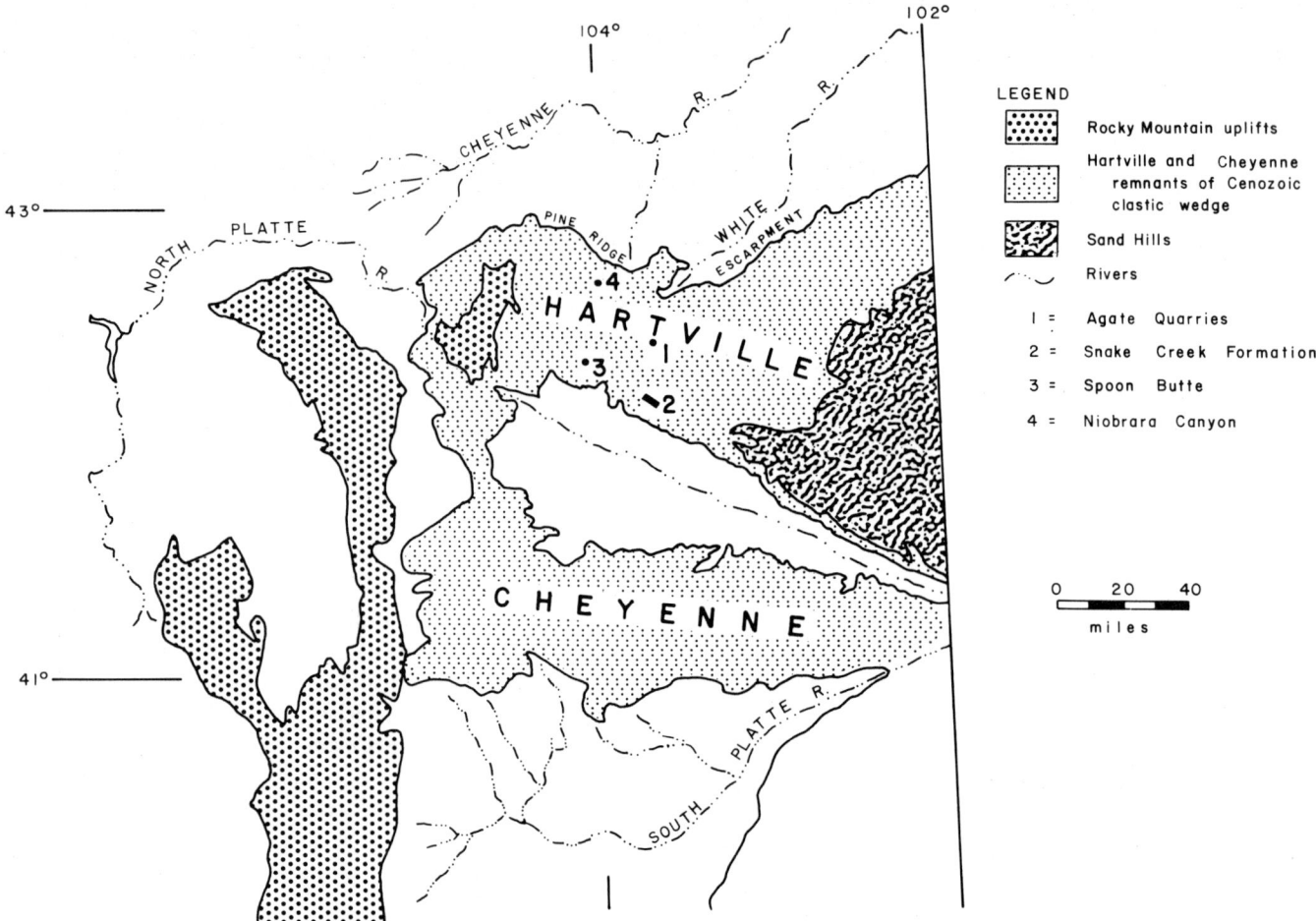

Figure 1. East of the Laramie Range of southeastern Wyoming, the Hartville and Cheyenne tablelands persist today as remnants of a thick Cenozoic clastic wedge that developed during the Oligocene and Miocene, eventually nearly burying adjacent uplift cores to the west, and thinning eastward into the midcontinent. The Hartville Table west of the Agate Quarries is primarily built by fine-grained volcaniclastic sediments of the White River and Arikaree Groups, and contains an abundant land mammal fauna. Late Miocene fluvial deposits of the Snake Creek Formation are the youngest Tertiary sediments aggrading the surface of the table, followed by the initiation of major dissection in the Plio-Pleistocene. The surface of the western Hartville Table (west of the Agate Quarries) is covered by upper Arikaree rocks. The stratigraphically highest Arikaree formation-rank unit (Upper Harrison of Peterson, 1907, 1909) is exposed in lateral continuity of outcrop over 3,100 km^2 (1,200 mi^2) of the table, and is well suited to determination of facies relations and reconstruction of Arikaree paleoenvironments. Cross sections of Figures 4 and 7 transect the Hartville Table, from the Pine Ridge and Niobrara Canyon to Agate Quarries, and demonstrate the geometry and facies of the Upper Harrison beds.

Field survey of the Upper Harrison beds exposed at the surface of the Hartville Table indicated that the principal lithologies were distributed among four major lithofacies (Fig. 5). Although much of the tableland is accessible only to offroad vehicles, all lithofacies can be readily observed in the center of the tableland at Agate Fossil Beds National Monument, where good exposures have been created by Quaternary incision of the Niobrara River. At Carnegie Hill, site of the main Agate Monument bone bed, these lithofacies occur one above the other in direct superposition (Fig. 6).

ARIKAREE FACIES AND PALEOENVIRONMENTS

Detailed descriptions of upper Arikaree lithofacies are presented elsewhere (Hunt, 1985). Here I briefly summarize distinguishing properties of each of the principal Upper Harrison lithofacies and associated biofacies and, from these observations, interpret the depositional environments.

A cross section of the northern Hartville Table from north to south (Fig. 7) allows an estimate of the relative volume of the various lithofacies of the formation. Massive, tuffaceous, silty sandstone with interspersed silica-cemented paleosols composes about 87 percent of the formation. Cross- and horizontally stratified fluvial deposits make up an additional 10 percent. Thin, widespread, tabular limestones contribute about 2 percent. Minor claystone lenses, diatomite, and volcanic tuffs constitute the remainder of the formation (1 percent).

Air-fall volcaniclastic loess with paleosols (Lithofacies F2-MS)

In outcrop, this widespread lithofacies is a texturally homogeneous tuffaceous silty sandstone without evident bedding (Fig. 8A). Only rare, relict, small-scale ripple lamination has been observed despite extensive outcrop exposure. In thin section (Fig. 8C), grains are angular to subangular, loosely packed, and sorting is good. Textural analysis of representative samples of this lithofacies (Fig. 9) indicates the uniformity of grain size, and the predominance of very fine-grained sand and silt.

Constituent grains (Tables 1 and 2) commonly include 30 to 54 percent volcanic glass shards, the remaining fraction about equally divided between quartz and feldspar. In addition to the glass shards, many of the framework grains appear to be of volcanic origin. Seventy-two percent of quartzose grains are strain-free monocrystalline quartz; resorption features and an occasional bipyramidal form are seen. The presence of glass-mantled and euhedral feldspars also suggests a volcanic source; sanidine is common, whereas microcline and perthite are very rare. When a plausible estimate of pyroclastic mineral grains is added to the

Figure 2. The western Hartville Table looking southwest toward the Laramie Range of southeastern Wyoming. The level surface is underlain by fine-grained volcaniclastic rocks of the Arikaree Group.

known shard content, the pyrogenic fraction of the rock commonly exceeds 50 percent and probably ranges from 50 to 75 percent.

In addition to its high pyroclastic content, these sandstones include a persistent epiclastic fraction, on the average no less than 12 percent (Table 1). Such grains tend to be somewhat more rounded, suggesting derivation from a separate source, most probably from the fine-grained bedload of ephemeral or intermittent streams. During arid intervals, these stream-transported epi-

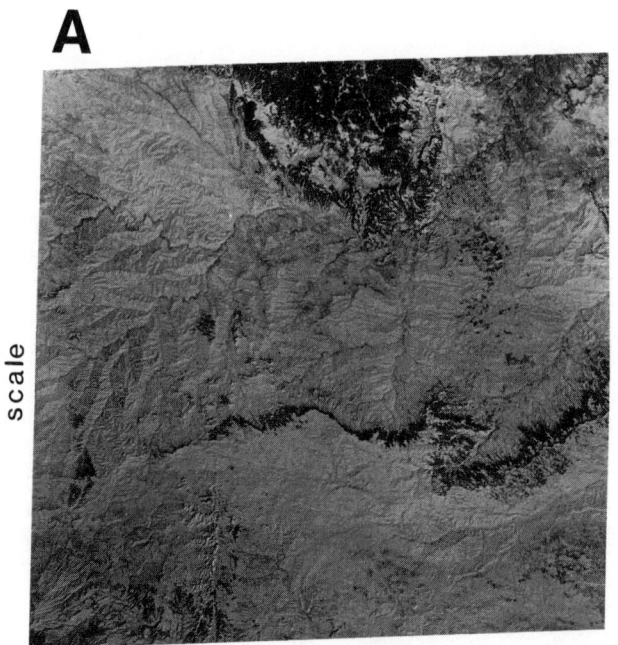
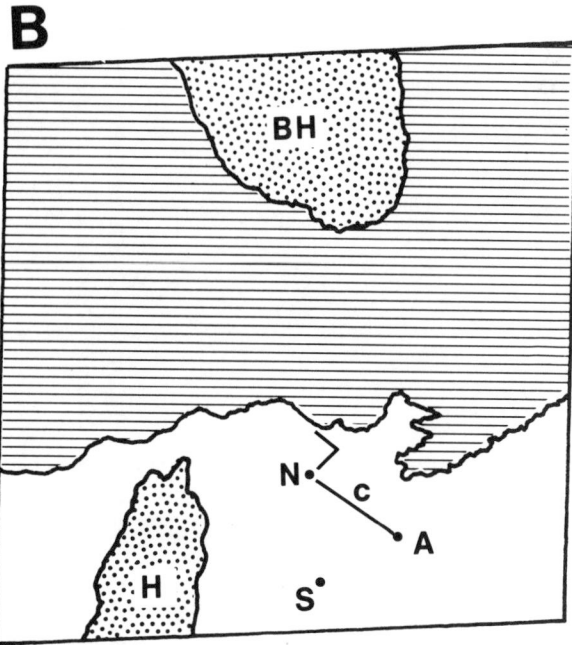

Figure 3. (A) Satellite photoimage and (B) diagram of the Hartville Table (white), Rocky Mountain uplifts (BH, Black Hills; H, Hartville Uplift), and Cheyenne–Hat Creek–White River lowland (horizontal line). The route of the cross section (c) of Figure 4 extends from the Pine Ridge escarpment (northern border of the Hartville Table) through the Niobrara Canyon (N) to Agate Fossil Beds National Monument (A). The Spoon Butte paleovalley (S) is situated about 24 km (15 mi) southwest of Agate.

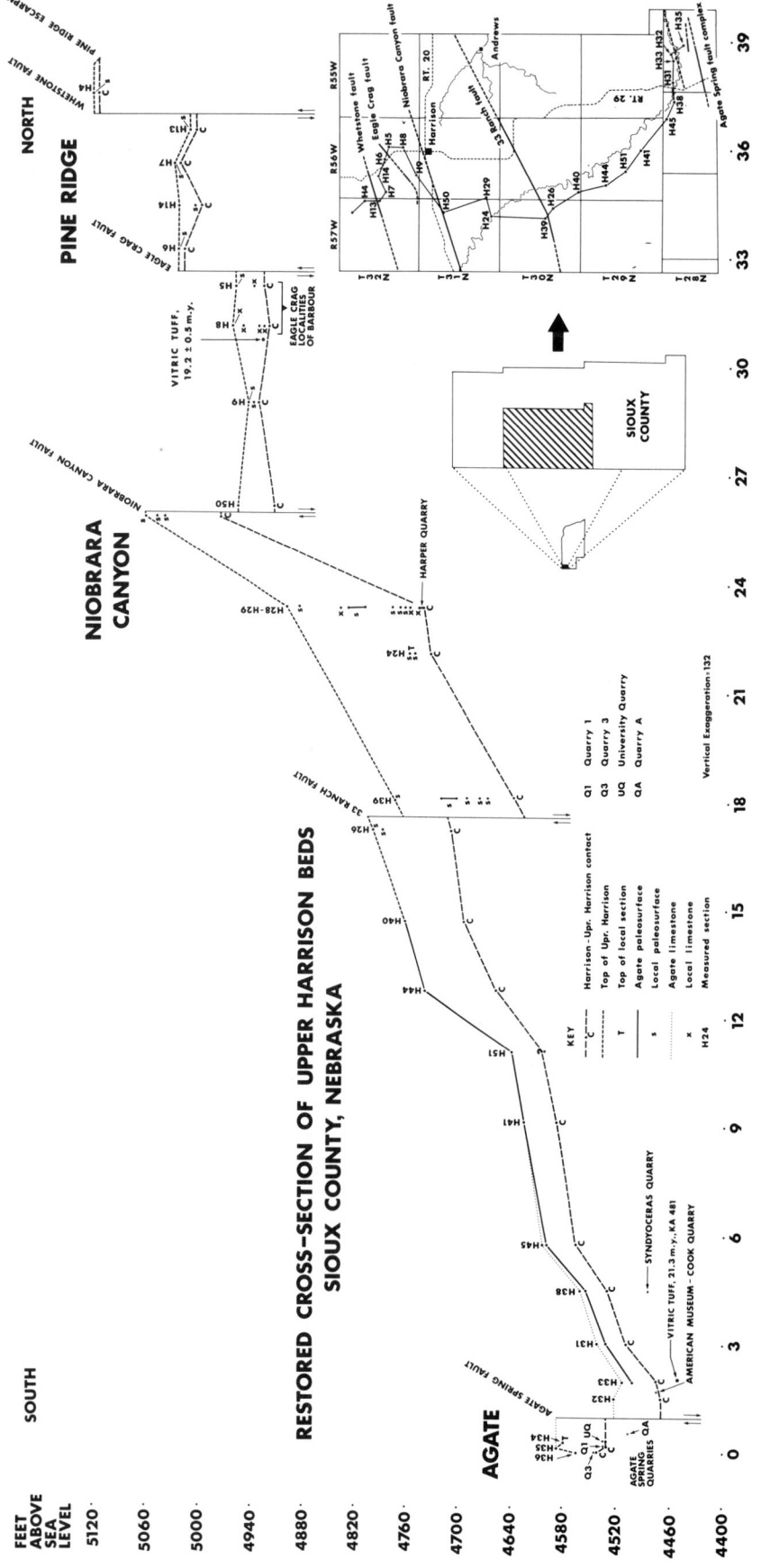

Figure 4. North-south cross section of the Upper Harrison beds, the youngest formation-rank rock unit of the Arikee Group in its stratotype area in the central Great Plains (modified from Hunt, 1978). The Upper Harrison beds form the surface of the western Hartville Table (see Fig. 3 for regional location of cross section). Lithofacies analysis of this cross section is shown in Figure 7.

PALEOENVIRONMENT	LITHOLOGY	LITHOFACIES	BIOTA
INTERCHANNEL PLAIN	TSS	Volcaniclastic loess with paleosols (F2-MS)	Mammals (chiefly ungulates), partially articulated to fully disarticulated, widely dispersed, attritional
BRAIDED STREAM	TSS, TST, CS, D, E	Fluvial channel and floodplain units (F1-XS) (minor eolian)	
EPHEMERAL LAKE	CM	Shallow holomictic lacustrine (F2-LS)	Aquatic invertebrates & plants
WATERHOLE	TSS/CT	Intertongued pond marl & fluvial sandstone (F1-CM)	Mammalian bone beds Aquatic invertebrates & plants Carnivore dens

Figure 5. Paleoenvironments of the Upper Harrison beds, Arikaree Group, interpreted from lithologies, lithofacies associations, and biota. Abbreviations: TSS = tuffaceous silty sandstone; TST = tuffaceous siltstone; CM = carbonate mudstone (micrite); CS = claystone; CT = calcareous tuff; D = diatomite; E = minor epiclastics and intraformational conglomerate.

TABLE 1. ESTIMATION OF PYROCLASTIC AND EPICLASTIC FRACTIONS OF VOLCANICLASTIC LOESS, UPPER HARRISON BEDS, ARIKAREE GROUP, USING AVERAGE PERCENTAGE VALUES FROM THE 11 SAMPLES OF TABLE 2

	Pyroclastic Only	Pyroclastic and/or Epiclastic	Epiclastic Only
Monocrystalline quartz (S)		18.6	
Monocrystalline quartz (U)			4.9
Polycrystalline quartz			1.6
Chert and chalcedony			0.6
Orthoclase feldspar			2.7
Perthite and microcline			0.5
Sanidine		3.5	
Plagioclase		12.6	
Plutonic rock fragments			1.4
Sedimentary rock fragments			0.2
Metamorphic rock fragments			0.1
Vitric shards	44.8		
Accessory minerals		5.4	
Totals	44.8	40.1	12.0

Note: 3.1 percent of grains are indeterminate or minor constituents.

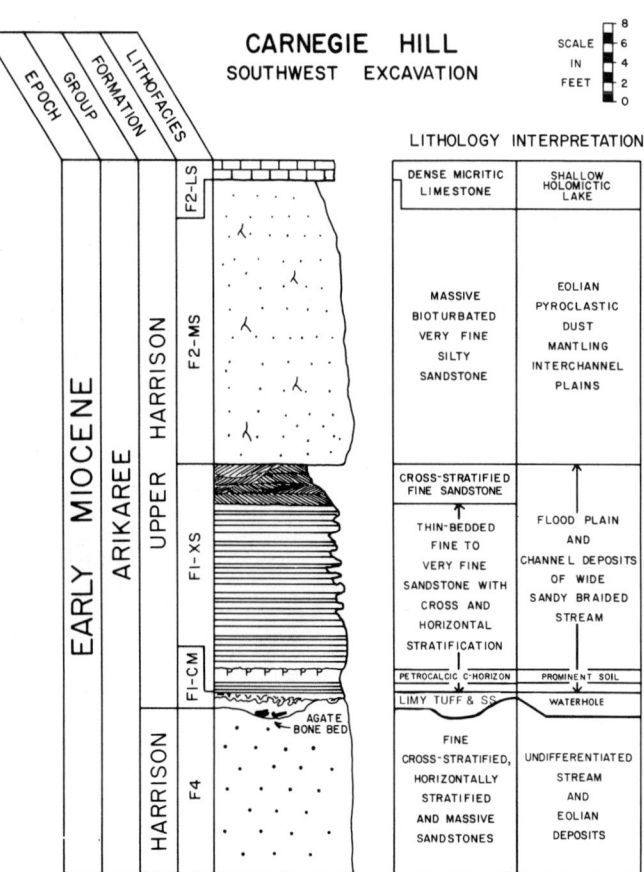

Figure 6. Lithofacies and paleoenvironments of the Upper Harrison beds at Carnegie Hill, Agate Fossil Beds National Monument, Sioux County, Nebraska. The principal Upper Harrison lithofacies of the Hartville Table are well developed within the national monument. The Agate bone bed occurs at the base of the formation in a sedimentation unit (lithofacies F1-CM) made up of intertongued calcareous tuff and tuffaceous fluvial sandstone that represents a waterhole environment.

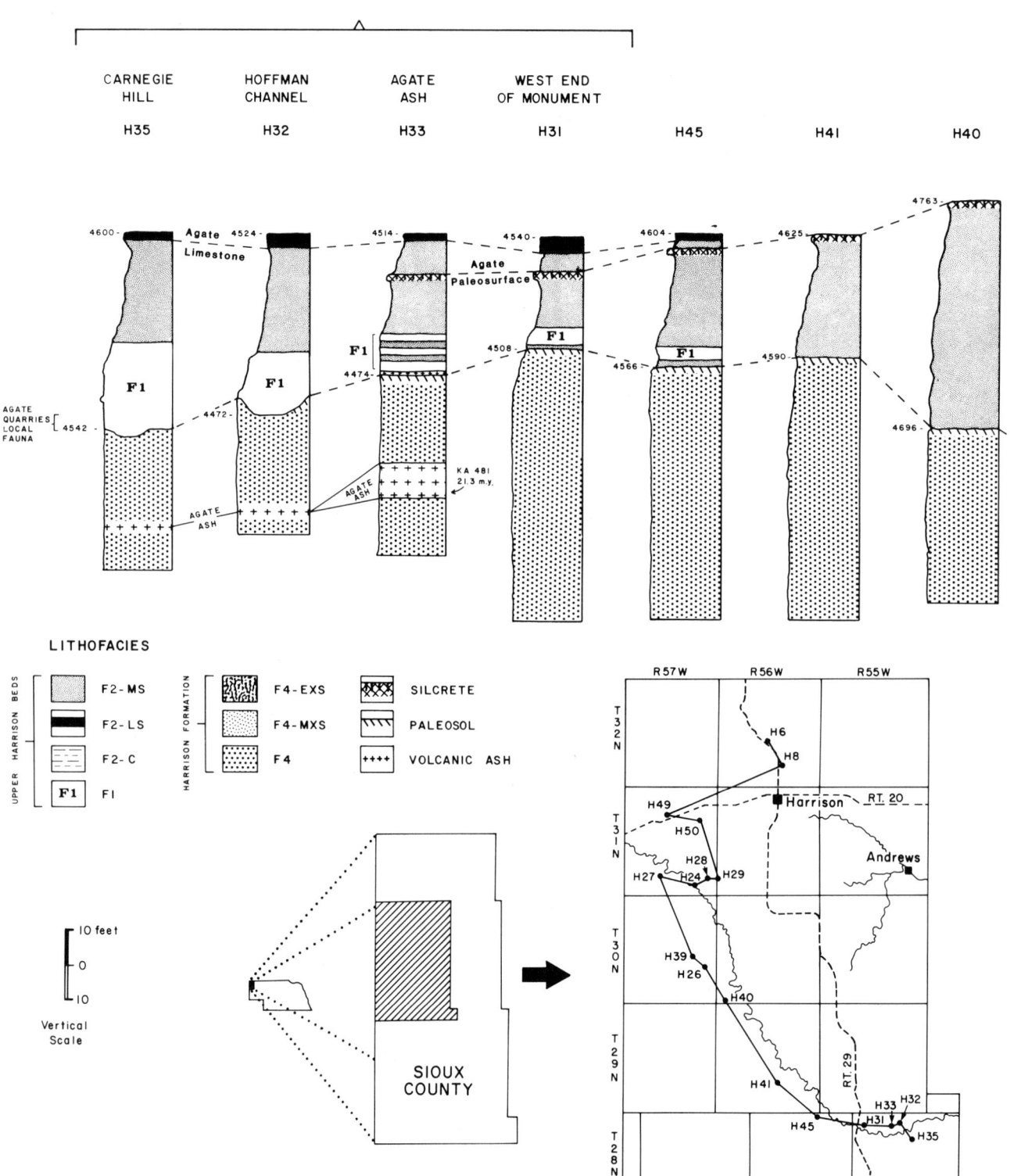

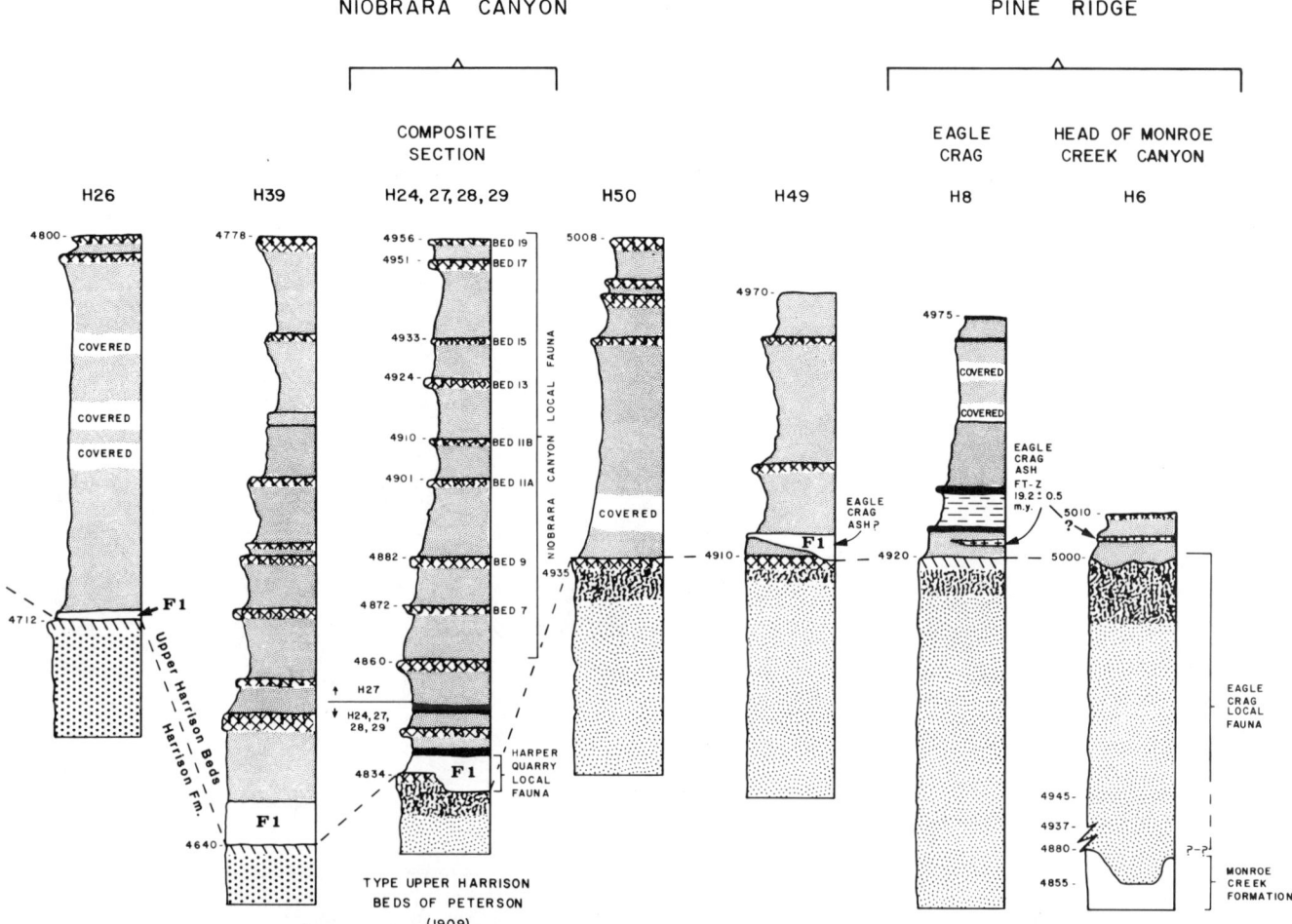

Figure 7. Stratigraphic profile of upper Arikaree sediments from the northern border of the Hartville Table (Pine Ridge) south to Agate National Monument (after Hunt, 1985). The Upper Harrison beds of Peterson (1907, 1909) are separated from the Harrison Formation of Hatcher (1902) by a regional disconformity. The stratigraphic position of key local faunas is indicated relative to lithofacies and dated volcanic tuffs. Silcrete paleosols in section H27 are numbered as marker horizons used in biostratigraphic sampling in the Niobrara Canyon. The Harrison Formation is divided into two facies (F4-EXS, F4-MXS) in the Niobrara Canyon and Pine Ridge areas, both eolian, but is undifferentiated (F4) south of the canyon. Elevations in feet above mean sea level.

clastic grains were mixed by wind with the dominant pyroclastic fraction. In Schmid's (1981) recent classification of mixed pyroclastic-epiclastic rocks, these Upper Harrison sandstones are tuffites (tuffaceous sandstones), defined by a pyroclast content of 25 to 75 percent relative to the epiclastic fraction.

At various levels throughout the vertical extent of these tuffaceous sandstones are uniformly flat indurated benches, cemented by silica and rich in rhizoliths, that represent paleosol horizons (Figs. 8A and B). Some can be traced for many miles as level, regional, geomorphic surfaces (Fig. 10), confirming the absence of paleotopographic relief. These surfaces dip gently to the southeast, defining a regional paleoslope that appears to be the result of their original early Miocene inclination, combined with a small amount of late Cenozoic tectonic tilt derived from the Hartville uplift–Black Hills structural axis (this axis forms the southeastern margin of the Powder River Basin). The uppermost 10 to 20 cm of these paleosols are well cemented by montmorillonite and opaline silica so that almost all pore space is filled, creating a silcrete duricrust: the amount of cement gradually diminishes downward until the parent sandstone is friable. Up to ten paleosols can be observed in superposition in some localities (Fig. 7, Niobrara Canyon).

Dense concentrations of siliceous rhizoliths and burrows are characteristic of these paleosols (Fig. 8A). In the more strongly developed paleosols, a laminar petrocalcic horizon is sometimes present in the uppermost meter, and the zone of highly concentrated roots and burrows that extends downward from the surface of the silcrete duricrust may attain a thickness of 1.8 m (6 ft). In all paleosols, the density of rhizoliths and burrows diminishes downward until the parent rock is devoid of biogenic structures.

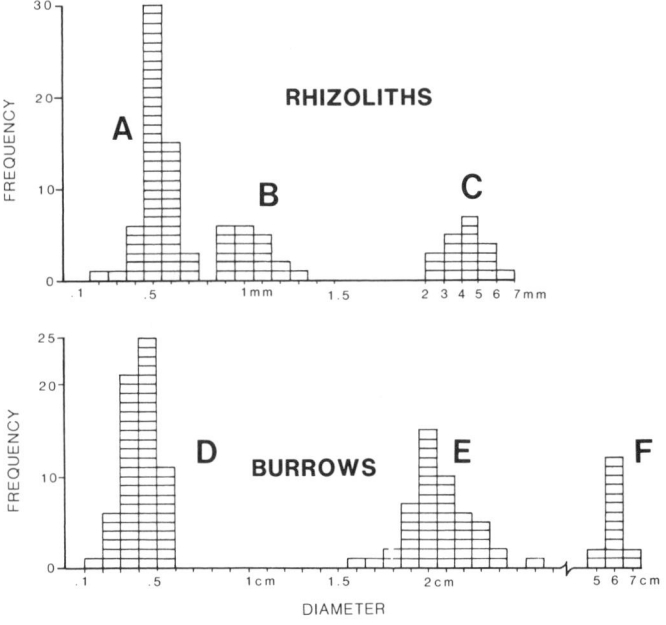

Figure 8. (A) Tuffaceous silty sandstone, massive, interpreted as an air-fall volcaniclastic loess (lithofacies F2-MS), with silcrete paleosol (ps) at top of section; zone (br) of rhizoliths and invertebrate burrows diminishing downward beneath silcrete (cliff height about 4.5 m; 15 ft). (B) SEM photomicrograph showing paleosol fabric made up of silt and very fine sand grains cemented by authigenic silica derived from pedogenic dissolution of volcanic glass shards and unstable mineral grains in the parent sandstone. Fine-diameter (0.5 mm) rhizoliths (rz) form dense networks in these paleosols, and are hollow tubes surrounded by a siliceous rind. (C) Photomicrograph showing well-sorted, very fine sand and silt of volcaniclastic loess lithofacies. Note numerous angular and subangular grains, many of which are volcanic glass shards. (D) Representative diameters of rhizoliths and burrows common to interchannel environments of the Upper Harrison beds are identified in paleosol (Agate paleosurface) of Figure 8A. A, grass root networks; B, isolated vertical roots; C, tap roots; D, sand-filled insect burrows; E, cylindrical burrow networks, similar to feeding burrows of the extant dung beetle *Dichotomius*; F, vertebrate burrows, probably rodents.

Measurement of a representative sample of burrow and rhizolith diameters found in the Agate paleosurface (Figs. 4 and 7), a well-developed paleosol at Agate Fossil Beds National Monument, illustrates the limited range of root and burrow types seen in these silcretes (Fig. 8D). Particularly common are dense anastomosed networks of tiny rhizoliths having a modal diameter of 0.5 mm (Fig. 8B). Less common are a group of somewhat larger vertical rhizoliths averaging about 1 mm in diameter, and a third group of taproot-like rhizoliths having diameters from about 2 to 7 mm.

Accompanying the rhizoliths are three distinct types of animal burrows (Fig. 8D): (1) small, somewhat tapering burrows with modal diameters of about 4 to 5 mm that are filled with packed sand, and may have been used by bees or wasps; (2) large cylindrical burrows with a modal diameter of about 2 cm that join at angles to form complex networks that show a striking similarity to the feeding burrows of the living dung beetle *Dichotomius* (Halffter and Matthews, 1966); and (3) small, sinuous vertebrate burrows ranging in diameter from 5 to 7 cm, probably made by small geomyoid rodents. Occasionally, large vertebrate burrows (Fig. 11) up to a meter in diameter and several meters in length, capable of housing mid-sized to large Carnivora, occur in association with these paleosols.

Within the lithofacies, the paleosol horizons are most strongly affected by diagenesis, which takes the form of slight alteration and dissolution of glass shards and unstable mineral grains. Silica mobilized during pedogenesis is converted to pore-filling cements and authigenic montmorillonite clay coatings on grains. Within these coatings, particularly at the deeper levels of the profile, much glass is generally fresh and not devitrified.

Associated with this lithofacies is an oreodont-camelid biofacies in which skeletal remains of these two ungulate groups

TABLE 2. PETROGRAPHIC COMPOSITION OF THE VOLCANICLASTIC LOESS LITHOFACIES, UPPER HARRISON BEDS, ARIKAREE GROUP, WESTERN NEBRASKA, BASED ON 11 THIN SECTIONS FROM NIOBRARA CANYON AND AGATE FOSSIL BEDS

Sample Number	H27-13 (%)	H27-15 (%)	H27-17 (%)	H27-18 (%)	H27-19 (%)	H27-20 (%)	H27-21 (%)	H32-9 (%)	H32-10 (%)	H32-12 (%)	H32-13A (%)	Average (%)
1. Total Quartzose	23	25	29	22	29	27	21	31	23	34	20	25.8
A. Monocrystalline Quartz (S)	20	16	19	15	23	21	15	22	15	27	12	18.6
B. Monocrystalline Quartz (U)	2	7	6	6	4	5	4	5	5	4	6	4.9
C. Polycrystalline Quartz	tr	2	3	1	1	1	1	3	3	2	1	1.6
D. Chert and Chalcedony	1	tr	1	0	1	tr	1	1	tr	1	1	0.6
2. Total Feldspar	20	18	22	19	25	22	20	23	25	26	19	21.7
A. Potassium Feldspar	7	5	5	5	9	4	7	10	9	8	4	6.6
a. Orthoclase	5	2	2	2	3	1	3	3	4	4	1	2.7
b. Perthite	0	0	0	1	tr	0	1	1	0	tr	0	0.3
c. Microcline	0	tr	0	0	0	0	0	1	tr	1	0	0.2
d. Sanidine	2	3	3	2	6	3	3	5	5	3	3	3.5
B. Plagioclase	9	12	16	11	14	13	10	13	15	15	11	12.6
C. Indeterminate Feldspar	4	1	1	3	2	5	3	0	1	3	4	2.4
3. Total Lithic	49	50	43	55	43	41	50	41	48	37	54	46.4
A. Igneous Rock Fragments	49	50	43	55	42	41	50	40	48	37	53	46.2
a. Plutonic (Microphaneritic)	1	2	2	1	1	3	1	1	tr	1	2	1.4
b. Volcanic (Nonvitric Igneous Aphanites)	0	0	0	0	0	0	0	0	0	0	0	0
c. Vitric (Shards)	48	48	41	54	41	38	49	39	48	36	51	44.8
B. Sedimentary Rock Fragments	0	tr	0	0	1	tr	0	0	0	0	1	0.2
C. Metamorphic Rock Fragments	tr	0	tr	0	0	0	0	1	tr	0	tr	0.1
4. Total Accessory Minerals	7	6	6	3	3	9	8	4	4	3	6	5.4
5. Other and Indeterminate	1	1	tr	1	tr	1	1	1	0	tr	1	0.7
6. Total All Grains (%)	100	100	100	100	100	100	100	100	100	100	100	
Total Grains Counted	270	331	339	234	326	223	229	335	326	340	285	

Abbreviations: S = straight extinction; U = undulose extinction; tr = less than 1 percent.

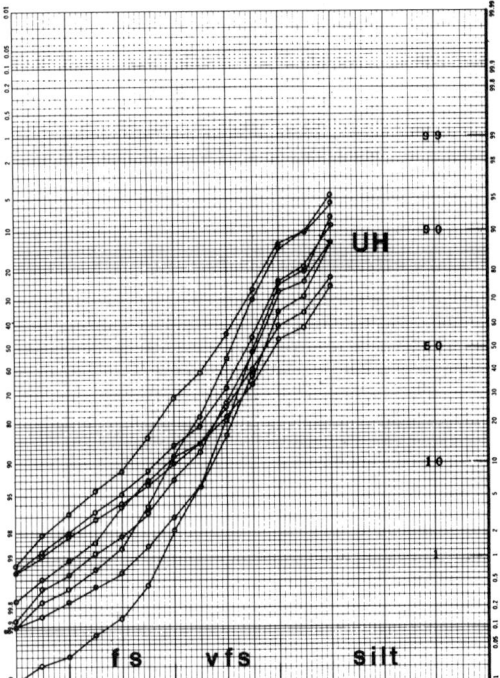

Figure 9. Textural analysis of representative samples (N = 8) of the Upper Harrison volcaniclastic loess lithofacies, the cumulative curves demonstrating predominance of very fine sand (vfs) and silt. X-axis in phi units; Y-axis in percent.

predominate. The most common oreodont is the small *Merychyus minimus,* present in large numbers at some localities. Partial skeletons of this species are widely scattered horizontally and vertically in outcrop, and are not concentrated in bone beds. At least two larger oreodonts (*Merycochoerus, Phenacocoelus*) also occur, one of which (*Merycochoerus matthewi*) is an important biostratigraphic indicator, and to date is only known from the upper part of the formation where nearly complete skeletons have been found cemented into the silcrete duricrusts. Camelids include the llama-sized *Oxydactylus* and *Tanymykter,* the smaller long-snouted *Michenia,* and the little gazelle-camel *Stenomylus.* Less frequently encountered members of the biofacies are a variety of horses (an anchithere and three species of *Parahippus*), the small rhinoceros *Menoceras,* several kinds of small deer-like ruminants, two species of peccaries, a protoceratid antelope, hypertragulid, chalicothere, large entelodont, tapir, lagomorph, several small rodents, and four families (canids, mustelids, amphicyonids, ursids) of carnivorans. At least 30 species of mammals occur in the biofacies.

Mode of occurrence of fossil mammals provides important information on environment of deposition. Skeletons most often occur as partially articulated remains (Table 3), generally with many bones missing, and show evidence of subaerial weathering. Each individual is isolated from the next, and the distribution of individual animals occurs at random horizontally and vertically in outcrop. Breakage patterns and bite marks on skeletal elements

Figure 10. Silica-cemented paleosols punctuate the volcaniclastic loess lithofacies, forming level regional surfaces (at arrows) that indicate early Miocene paleotopography. These surfaces are interpreted as flat interchannel plains largely populated by grasses, and are prominently exposed where the Niobrara Canyon incises the Hartville Table.

Figure 11. Paleosol horizons in the upper Arikaree Group may include large-diameter linear burrows (open arrow) interpreted as shelters excavated by mid-sized to large Miocene Carnivora. At Agate National Monument, skeletons representing five species of the Upper Harrison carnivore community (amphicyonids, mustelids, canid) have been found within such burrows. These burrows may extend to 1.8 m (6 ft) beneath the paleosol, indicating a water table below this level and a well-drained soil profile.

TABLE 3. DEGREE OF ASSOCIATION OF CRANIAL WITH POSTCRANIAL REMAINS IN THE OREODONTS *MERYCHYUS MINIMUS* AND *MERYCOCHOERUS MATTHEWI*, VOLCANICLASTIC LOESS LITHOFACIES, NIOBRARA CANYON*

	N	%
Merychyus minimus		
Association of		
Skull, jaw, and postcranials	10	25
Skull and postcranials	12	30
Jaw and postcranials	7	17.5
Skull and jaws	5	12.5
Skull only	4	10
Jaw only	2	5
Totals	40	100
Merycochoerus matthewi		
Association of		
Skull, jaw, and postcranials	3	60
Skull and jaw	1	20
Postcranials	1	20
Totals	5	100

*The *Merychyus* sample was collected in 1901 by Peterson for the Carnegie Museum, Pittsburgh. The *Merycochoerus* sample includes all known specimens from the canyon.

demonstrate extensive scavenging and carnivory prior to sediment burial. The commonly encountered stages of dental eruption and wear in jaws show most individuals to be either very young or quite old, with few prime adults. Bones are found in the cemented paleosols and the more friable sediments between the paleosols. Remains of ungulates predominate in this lithofacies (at least 20 species); carnivores are much less common.

Interpretation. Loess commonly occurs as regionally extensive sheets, often several hundred feet in thickness, with a terrestrial fauna, absence of stratification, lack of consolidation, and a modal particle size in the silt category with no grains coarser than fine sand (Reineck and Singh, 1980; Pye, 1987). Paleosols may be present throughout thick loessic sequences (Burbank and Li Jijun, 1985). Air-fall pyroclastic debris, when deposited in terrestrial environments under similar climatic conditions, can exhibit loess-like features, blanketing extensive geographic areas as it settles from the air. In grassland or savanna-parkland, fine-grained pyroclastics periodically falling on a subdued topography of minimal relief, stabilized by evenly distributed vegetation, could contribute to the development of a broad, level regional geomorphic surface. Grasses and other low-growing plants will hold the sediment and, if plant growth keeps pace with the influx of wind-transported sediment, maintain the surface over time. Because of the trapping effect of the vegetation, these loessic deposits will lack stratification (Reineck and Singh, 1980, p. 237).

Arikaree volcaniclastic loess is the product of eolian deposition, punctuated by flat, regionally extensive paleosols that indicate periodic stabilization of the land surface during long pauses in accumulation of pyroclastic debris on level plains. Based on the dominance of uniformly fine-diameter rhizolith networks, these plains were covered with closely spaced, finely rooted vegetation, probably grasses. Absence of any large-diameter rhizoliths or silicified wood suggests either the absence of trees and woody brush, or the lack of preservation of woody tissues. Growth of vegetation probably easily kept pace with loess deposition, based on the low sediment accumulation rates (see below) calculated for these rocks. The angular, well-sorted, uniformly fine-grained texture of the sediment, its high pyrogenic content, and its blanketing regional geometry, demonstrate a pyroclastic eolian origin. Lack of evident stratification in the sediment, angularity of grains, and the open grain packing suggest little or no postdepositional modification by fluvial processes or compaction of air-fall sediment, except for displacement resulting from the activity of burrowing animals and plants. However, transport of fine pyroclastic detritus into the North American midcontinent may have been a multistage process in which primary air-fall deposits were progressively redistributed farther eastward by prevailing westerly winds through the agency of seasonal dust storms, in a manner similar to the desert loess of China (Liu, 1985).

A limited number of kinds of rhizoliths, all of small diameter, the predominance of very fine-diameter root casts interpreted as grasses, and the constancy of a few basic burrow types in these fossil soils suggest that a uniform low-growing vegetation and stable, well-established soil fauna were widely distributed over broad plains between the stream courses. Absence of significant diagenesis of unstable glass shards and pyrogenic mineral grains, except within paleosols, and the leaching of calcium carbonate from the soil profile into shallow laminar petrocalcic horizons less than a meter below the ground surface, would argue for a dry climate in which limited rainfall was probably seasonally distributed. In such a semiarid setting within the continental interior, it seems likely that rainfall achieved only a limited penetration of the soil. Carbonic acid in rainwater, together with pedogenic plant and animal activity, dissolved silica from shards and unstable minerals, coating grains in the zone of aeration with montmorillonite clay, thereby reducing porosity. Eventual filling of remaining pore spaces with opaline silica cement gradually converted the land surface to a cemented duricrust. The pale brown tones (10YR 6.5-7/2-4) of this lithofacies suggest the presence of dehydrated iron oxides developed in well-drained interchannel areas. There oxidation prevailed in the zone of aeration, due to a low water table evidenced by the considerable depth of penetration of burrow systems beneath the paleosol surfaces.

The mammal fauna, dominated by diverse species of herbivorous ungulates, argues for a reasonably diverse flora occupying the region. However, many of the less common ungulates found in this lithofacies may have ranged only infrequently into the interchannel reaches, or maintained only small populations in that environment. This is supported by the observation that bones from some of the larger ungulates (chalicothere, entelodont, tapir) are scarcer in this lithofacies than bones from smaller, more common species (oreodont *Merychyus,* camelids). Because larger bones commonly withstand environmental processing better than smaller bones, the rarity of these large ungulates in interchannel deposits probably actually reflects habitat preferences and/or relative population sizes.

Based on evaluation of their dentitions, both browsers (anchithere, chalicothere) and grazers (gazelle-camel) are included in the fauna. Some species (deer-like ruminants and the hypertragulid) almost certainly lived in well-vegetated settings, indicating the presence of tall grasses and/or low-growing shrubs and brush that served as both food resource and protective cover. Probably the browsers, such as tapir, chalicothere, and small ruminants, which are not as commonly preserved as oreodonts and camels, lived in more heavily vegetated settings such as stream-border savanna woodland, and may have only occasionally penetrated the more open interchannel environment. It seems likely that oreodonts and camels were either grazers or generalists, able to maintain larger populations in the interchannel environment, much as large numbers of guanaco are known to have populated the South American grasslands (Hatcher, 1903).

The absence of any fully arboreal mammals indicates woodland was not a common or widespread habitat type. Although many of these ungulates appear to be adequately adapted to open environments such as grassland, or to sparsely wooded savanna parkland or woodland, the rhizolith assemblage of the paleosols seems to reflect a treeless savanna with, at most, scattered brush in the interchannel environment. More densely vegetated settings were limited to the streams, where woody vegetation, low shrubs, and tall grasses probably prevailed.

Mammalian skeletons accumulated in the parent loess and paleosols as a time-averaged sample through normal attritional death events. In accord with this interpretation are the attritional age profiles of ungulates, prevalence of scavenged and subaerially weathered skeletons, the greater numbers of ungulates relative to carnivores approximating normal predator/prey ratios, and the spatial distribution of skeletons randomly scattered through the loess. Partial articulation of many fossil skeletons suggests drying (mummification) of carcasses in a semiarid or arid environment, following initial dispersal of some parts of the animal by scavenging.

The pattern of sedimentation of volcaniclastic loess is relevant to the mode of preservation of mammal skeletons in these rocks. Differing rates of sedimentation and the amount of sediment supplied at any given time within a particular climatic context are important influences on preservation of skeletal material. An average sediment accumulation rate for the Arikaree Group in its type area in western Nebraska can be derived by comparing its thickness (about 213 m or 700 ft) to the amount of time bracketed by dated vitric tuffs at the base and top of the section. The base of the Arikaree Group has been dated at 27 to 29 Ma (Obradovich and others, 1973; Naeser and others, 1980; Tedford and others, 1987, p. 197) using both the fission-track method on zircon and potassium-argon dating of biotite. The upper part of the Arikaree Group has been dated at 19.2 ± 0.5 Ma using the fission-track method on zircon (Hunt and others, 1983). Allowing 8 to 10 m.y. for deposition of the Arikaree Group, we can calculate an average sediment accumulation rate of 2 to 2.5 cm (0.8 to 1 in) per 1,000 yr.

Accumulation of fine-grained Arikaree sediment was certainly episodic. Although this is evident when one considers the paleosol-punctuated Upper Harrison beds, documented intervals of nondeposition within loessic facies of the Arikaree at lower stratigraphic levels are more difficult to discern. Yet the existence of such horizons is demonstrated by the staggered vertical distribution of vertebrate burrow fills scattered through nearly 45 m

Figure 12. Thin tabular beds of white siliceous limestone with a biota of ostracods, aquatic pulmonate gastropods, and plant rhizoliths are interpreted as shallow, ephemeral, holomictic lakes. These limestones contain no aquatic vertebrates, such as fish, nor any mammal remains. No streams enter or exit these lakes; they appear to be isolated, filling local topographic depressions in the level surface of interchannel plains. Drying of the lake muds is indicated by desiccation polygons, and overprinting of the mudstone by soil features.

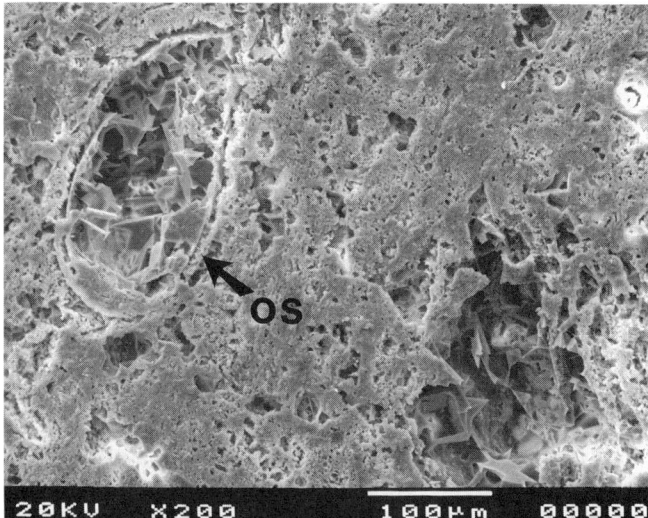

Figure 13. SEM photomicrographs of silica fabric in thin micritic limestones deposited in shallow, ephemeral, holomictic lakes. All calcite has been removed by treatment with dilute hydrochloric acid. Silica cement is so pervasive that removal of calcite leaves intact a cohesive siliceous framework that preserves the form of the rock. Top: residual fabric of diagenetic silica that surrounds micrite crystals, spar calcite, and ostracods (os). Bottom: detailed view of opaline silica cement surrounding cavities (m) once occupied by micrite crystals prior to acid treatment.

(150 ft) of the Arikaree Group beneath the Upper Harrison beds. These burrows were excavated by the castorid *Paleocastor,* an extinct, gopher-like rodent found on occasion within the helical burrow spiral or within the burrow's nesting chamber (Peterson, 1905). The burrows occur at various stratigraphic levels throughout volcaniclastic loess and stratified eolian lithofacies of the upper Monroe Creek and Harrison Formations where prominent paleosols are absent or so weakly developed as to be unidentifiable. Each burrow mouth marks a horizon where deposition was interrupted by a temporal interval long enough to allow burrow construction and occupation of a land surface. The fact that burrows occur at multiple levels attests to the episodic nature of sedimentation throughout this loessic sequence.

Thus, considering the low average sediment accumulation rate calculated for the Arikaree Group, and the evidence for episodic aggradation, thin sediment increments within friable massive loess may contain skeletal material accumulated over long periods of time. And if, in fact, there were brief periods when considerable thicknesses (>10 cm) of volcaniclastic loess were deposited, there also must have been very long intervals when no accumulation took place. When seen in the field, the towering cliffs of Arikaree rock present an imposing record of volcanic sedimentation, but the actual rate of accumulation of this predominantly loessic material was probably so slow that it was hardly discernible in terms of the human lifespan.

Lacustrine carbonate mudstone (Lithofacies F2–LS)

Thin beds of white siliceous limestone are occasionally found in association with the volcaniclastic loess lithofacies (Fig. 12). These limestones are texturally homogeneous micrites, often extensively pervaded by silica (Fig. 13) in the form of chert. At several localities where primary structure can be observed, the limestones are thinly laminated. Each occurs as a thin widespread sheet, commonly from 3 to 30 cm thick (0.1 to 1 ft, maximum of 1.8 m or 6 ft), and as much as 5 mi^2 in area, grading at its margins into calcareous silty sandstone and then volcaniclastic loess. Only at the lake margin where limestone grades into calcareous sandstone is there a detrital fraction, consisting of rounded silt and sand grains mixed with micrite, or thin detrital laminae of silt and very fine sand interbedded with thin, mud-cracked clay laminae.

The upper surface of these limestones is sometimes burrowed, invaded by rhizoliths, and may exhibit brecciation indicative of pedogenesis. Occasional desiccation cracks demonstrate fluctuating water levels and at least local subaerial exposure. Desiccation polygons to 40 cm in diameter have been measured. The upper and lower bounding surfaces are sharp, and the limestone does not grade into the sandstones above and below, although leaching may carry carbonate downward, creating a calcite pore cement in subjacent sandstone that mimics a gradational lower contact.

If the limestone is dissolved in dilute hydrochloric acid, removal of calcite leaves intact a cohesive siliceous fabric that preserves the external form of the rock but is much lighter in weight. SEM photomicrographs show that this siliceous matrix completely encloses micrite grains, spar, and the calcareous shells of ostracods (Fig. 13). X-ray diffraction indicates that the matrix is opaline silica (opal-CT).

These carbonate mudstones contain a depauperate biota of aquatic invertebrates and plants. Local concentrations of aquatic pulmonate gastropods (Lymnaeidae) and freshwater ostracods are represented by primary shell material as well as shells replaced by spar calcite (Fig. 13). Plant roots of small diameter are preserved within the micrite mud, including some with original cell structure intact and apparently silicified. Invertebrate burrows up to 1.5 cm in diameter penetrate downward into the mudstone as vertical tubes filled with micrite. Significantly, no aquatic vertebrates such as fish, turtles, or frogs are found in these mudstones.

Interpretation. The common association of thin, sheet-like, finely laminated carbonate mudstones with volcaniclastic loess; their freshwater biota of ostracods, pulmonate snails, and plant debris; and the evidence of drying and pedogenetic overprinting of these muds, combine to indicate shallow, ephemeral, holomictic lakes scattered over level terrain east of the Hartville uplift. Lack of significant amounts of detrital grains in the mudstones, except along lake margins where wave action introduced a small detrital fraction, demonstrates isolation of the lakes from local streams and suggests that littoral vegetation prevented significant influx of wind-transported sediment.

Water-level fluctuations and eventual drying subjected the lake muds to subaerial exposure, resulting in burrowing and fissuring of the surface, later followed in some locations by brecciation and recementation. Over time, silica-rich waters (probably from devitrification of ash) in association with pedogenesis infiltrated the initially porous micrite, resulting in growth of authigenic opaline cements that formed a pervasive rock fabric.

Similar micritic limestones that formed in shallow ponds and lakes have been reported in the Paleogene of southern France by Freytet and Plaziat (1982). In their monographic discussion of nonmarine carbonates in Languedoc province, they describe lacustrine limestones that developed in association with Eocene molassic sedimentation resulting from elevation of the Pyrenees.

About 90 percent of the section is contributed by detrital sediments, about 10 percent by lacustrine and paludal carbonates. Abundant calcareous soil features and lacustrine limestones within the molasse sequences were attributed to "the combined effects of a carbonate substratum and relative aridity of the paleoclimate."

Within the Languedoc molasse, limestones formed in ephemeral ponds and much larger flood-plain lakes, and in swamps marginal to these lakes. Freytet and Plaziat (1982) identified thin lenses (about 3 to 30 cm) of dark Eocene micrites containing charophytes, *Lymnaea,* and planorbid snails as ephemeral pond deposits. Lack of detrital sediment in these micrites from the surrounding flood-plain environment was attributed to the filtering effect of vegetation at the margin of the pond. In addition to these ephemeral flood-plain ponds, Freytet and Plaziat (1982) described a second class of granular and micritic limestones several meters in thickness and several tens of kilometers in areal extent, hence similar in geometry to Upper Harrison lacustrine limestones, but formed in even larger lakes on fluvial plains and associated with both fluvial and marsh sediments: these lakes "seem to have been holomictic, being relatively shallow . . . with respect to their width . . . ." Charophytes are often abundant and fish are absent.

Because the Upper Harrison limestones generally are isolated bodies within the volcaniclastic loess lithofacies, they differ to some degree in environmental setting from ephemeral pond and lacustrine limestones deposited on the Eocene flood plains of Languedoc. However, the Upper Harrison Miocene limestones of the Hartville Table share many similarities with these Eocene lacustrine units, namely: (1) generally thin, geographically extensive beds, indicating shallow widespread bodies of standing water; (2) homogeneous micrite fill deposited in a shallow holomictic lacustrine setting; (3) a depauperate aquatic biota, largely limited to pulmonate gastropods, charophytes, and ostracods; (4) absence of aquatic vertebrates, particularly fish; (5) paucity of detrital sediment except at periphery of lake, suggesting negligible fluvial input and a curtain of perilacustrine vegetation; (6) desiccation and pedogenesis of emergent surfaces of lacustrine units, indicating alternating wet and dry intervals; (7) diagenetic replacement of carbonate by silica.

Fluvial volcaniclastic sandstone (Lithofacies F1–XS)

About 10 percent of the formation is represented by horizontally and cross-stratified tuffaceous silty sandstones, texturally and mineralogically similar to the massive tuffaceous silty sandstones of the volcaniclastic loess lithofacies, but containing a somewhat greater percentage of epiclastic grains. These stratified volcaniclastics are restricted to wide, relatively shallow paleovalleys cut into older Arikaree rocks (Fig. 14). Thin- to thick-bedded, lower-flow-regime, silty, very fine-grained sandstones form the bulk of these sediments; upper-flow-regime deposits appear to be restricted to the basal fill of large channel scours. Two major paleovalley segments (Agate Monument and Spoon

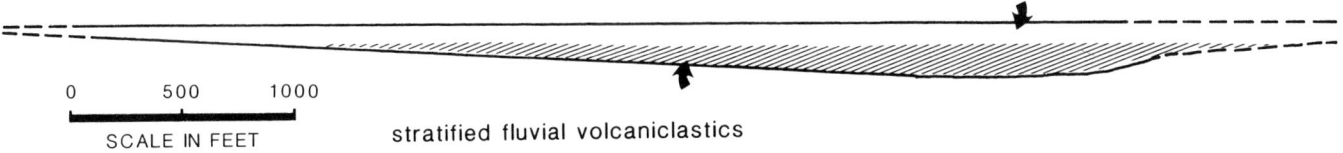

Figure 14. The cross-sectional geometry of the Spoon Butte paleovalley in southeastern Wyoming is typical of the wide, shallowly incised Upper Harrison valleys of the western Hartville Table. Fluvial stratified volcaniclastics (with minor epiclastic prisms) predominate in the lower part of these valley fills, whereas the upper part is dominated by massive fine-grained volcaniclastic loess. Mammalian fossils in these paleovalleys commonly occur as scattered skeletal debris derived from attritional processes, and are most abundant in the lower fluvial part of the fill. No vertical exaggeration.

Butte paleovalleys) filled with Upper Harrison sediments have been identified east of the Hartville Uplift where they incise the surface of the Hartville Table. The uplift core includes Precambrian granitic and metamorphic rocks, associated with a flanking platform of Paleozoic carbonates.

At Agate Fossil Beds National Monument, an axial valley filled with fine-grained fluvial volcaniclastics, trends southwest-to-northeast for about 6.5 to 8 km (4 to 5 mi). In the Agate Monument area (Fig. 6), these stratified volcaniclastics do not exceed 9 m (30 ft) in thickness, whereas their lateral extent in a cross-sectional profile of the paleovalley attains 6.5 km (4 mi). Along the southern margin of this valley, the stratified volcaniclastics abruptly terminate against low, vertical stream banks about 1.8 m (6 ft) in height. Large, angular blocks of untransported lithified Harrison sandstone derived from bank collapse occur along the stream margin. The northern margin of the paleovalley is identified by intertonguing of stream deposits (stratified volcaniclastics) with interchannel sediments (massive volcaniclastic loess) in the absence of a defined stream bank.

Sediments filling the Agate Monument paleovalley have been described by Vicars and Breyer (1981) and Hunt (1985). The lower part of the valley fill is formed by as much as 8 m (26 ft) of stratified volcaniclastics, the upper part by 9 m (30 ft) of massive volcaniclastic loess. The lower part is characterized by: (1) lower-flow-regime, thin- to medium-bedded, small-scale, cross-laminated and rippled sandstones, with occasional thin interbedded silt laminae, deposited in association with small shallow scours; and (2) larger-scale trough-cross-stratified sandstone resulting from migration of sinuous-crested bedforms. At Carnegie Hill, 78 percent of the stratified volcaniclastic fill belongs to this first category, and includes predominantly ripple-, low-angle cross-, trough-cross-, and horizontally laminated thin sandstones, indicative of shallow, laterally widespread sheetflow of low energy.

The large-scale trough cross-strata were viewed by Vicars and Breyer (1981) as the dominant bedform preserved within the Agate Monument paleovalley. The troughs average 3.5 m (11.5 ft) in width, 34 cm (1.1 ft) in height, and are grouped into cosets ranging in thickness from 0.6 to 4.3 m (2 to 14 ft). Each coset rests upon an erosional surface, with granule to pebble lag debris commonly concentrated at its base. At low-water stage, the surface of a coset was often subaerially exposed, and overprinted by soil development, so that the bedding in its upper part was obliterated by rhizoliths and burrows. These stacked cosets are on occasion interrupted by insertion of a bed of massive, fine-grained, volcaniclastic sandstone, which also may be overprinted by a rhizolith-burrow complex. Such beds, however, differ from the cosets in that they lack any evidence of primary stratification, and are believed to be eolian. Similarly, the thin-bedded cross-stratified and rippled deposits also occasionally contain thin beds of structureless sand, and are overprinted at various levels by rhizolith-burrow networks.

Median grain size of these deposits is very fine sand (3.6 phi), and the silt-clay content averages 17 percent. Textural variation is limited to very fine to fine-grained sandstone, siltstone, and locally derived lithic granule to pebble lag deposits at the base of sedimentation units. Petrographic analysis of these stratified volcaniclastics (Table 4) shows that their composition is not significantly different from the petrographic composition of the volcaniclastic loess (compare Table 2). The stratified volcaniclastics filling the Agate Monument paleovalley differ from the interchannel volcaniclastic loess in having about 5 percent more plagioclase and undulose monocrystalline quartz, about 5 percent less straight-extinction monocrystalline quartz, about 2 percent more chert, about 8 percent less volcanic shards, and 1.4 percent more sedimentary rock fragments. These differences indicate a small increase in the epiclastic fraction and slight decrease in the pyroclastic component relative to the loess. The pyroclastic fraction of these stratified deposits remains significant, a minimum of about 50 to 65 percent of grains.

At Spoon Butte about 29 km (18 mi) west of Agate, the thick sedimentary fill of the largest known Upper Harrison paleovalley is displayed in cross section along the west escarpment of the butte (Fig. 14). The axial valley fill also is preserved in outcrop both west and east of the butte, for a distance of at least 10 km (6 mi), where modern stream dissection has exposed about 68 vertical m (225 ft) of chiefly fine-grained volcaniclastic sediments, occupying a paleovalley about 3.2 km (2 mi) in width.

Stratified volcaniclastic sandstones and siltstones predominate in the lower 38 m (125 ft) of the fill (Fig. 15A); massive fine-grained volcaniclastic sandstones make up most of the upper 30 m (100 ft) of section. The range of sedimentary structures,

TABLE 4. PETROGRAPHIC COMPOSITION OF THE FLUVIAL VOLCANICLASTIC LITHOFACIES, UPPER HARRISON BEDS, ARIKAREE GROUP, BASED ON 11 THIN SECTIONS FROM THE AGATE PALEOVALLEY (AGATE FOSSIL BEDS NATIONAL MONUMENT), COMPARED WITH AVERAGE VALUES FOR 13 GRAIN MOUNTS (VERY FINE SAND FRACTION) FROM THE SPOON BUTTE PALEOVALLEY

Sample Number	H32-6 (%)	H32-7 (%)	H32-8 (%)	H32-11 (%)	H35-2 (%)	H35-3 (%)	H35-4 (%)	H35-5 (%)	H35-6 (%)	H35-7 (%)	H35-8 (%)	Average (Agate) (%)	Average (Spoon Butte) (%)
1. Total Quartzose	25	23	33	30	24	35	29	26	30	28	33	28.7	13.1
A. Monocrystalline Quartz (S)	19	16	24	18	5	13	11	12	16	9	8	13.7	7.8
B. Monocrystalline Quartz (U)	4	4	6	6	16	17	16	11	10	15	11	10.5	2.0
C. Polycrystalline Quartz	2	2	1	2	1	1	0	tr	1	0	1	1.0	tr
D. Chert and Chalcedony	tr	1	2	4	2	4	2	3	3	4	3	2.5	3.3
2. Total Feldspar	22	27	25	29	27	33	24	27	32	19	26	26.4	40.9
A. Potassium Feldspar	7	9	8	10	15	11	4	8	13	1	4	8.2	....
a. Orthoclase	1	6	3	2	....	....	....	....	....	....	....	3.0	....
b. Perthite	0	0	2	1	....	....	....	....	....	....	....	0.7	....
c. Microcline	1	1	0	1	....	....	....	....	....	....	....	0.7	1.4
d. Sanidine	5	2	3	6	....	....	....	....	....	....	....	4.0	....
B. Plagioclase	15	17	15	19	11	21	19	18	18	17	21	17.4	....
C. Indeterminate Feldspar	0	1	2	0	1	1	1	1	1	1	1	0.9	....
3. Total Lithic	49	48	35	34	45	28	43	41	32	48	37	40.0	42.4
A. Igneous Rock Fragments	49	48	35	33	40	24	40	38	32	47	35	38.3	39.8
a. Plutonic (Microphaneritic)	1	tr	2	2	2	2	1	0	2	1	1	1.3	3.8
b. Volcanic (Nonvitric Igneous Aphanites)	0	0	0	0	0	0	0	0	0	0	0	0	0
c. Vitric (Shards)	48	48	33	31	38	22	39	35	30	46	34	36.7	36.5
B. Sedimentary Rock Fragments	0	0	0	1	5	4	3	3	0	1	1	1.6	2.6
C. Metamorphic Rock Fragments	0	0	0	tr	0	0	tr	0	0	0	1	0.3	tr
4. Total Accessory Minerals	4	2	6	7	3	4	3	6	6	5	4	4.5	2.9
5. Other and Indeterminate	tr	tr	1	0	1	0	1	0	0	tr	0	0.3	....
6. Total All Grains (%)	100	100	100	100	100	100	100	100	100	100	100		
Total Grains Counted	332	333	329	333	356	327	336	340	288	305	339		

Abbreviations: S = straight extinction; U = undulose extinction; tr = less than 1 percent.

textures, and lithologies displayed in the lower 38 m (125 ft) of the fill is greater than that observed in the Agate Monument paleovalley. The Spoon Butte paleovalley also is more deeply incised than at Agate, and the stratified fill itself is thicker.

Coarsest sediments of the Spoon Butte paleovalley are lenses of intraformational conglomerate generally found at and near the base of the fill. They are scattered along the valley floor and sides as discontinuous conglomerate prisms, often displaying an elongate, asymmetric, bar-like form in plan view. A single clast type makes up the framework of this clast-supported conglomerate: as the paleovalley was progressively incised into subjacent Arikaree sandstone, the finer volcaniclastic sediment was carried away by stream currents while numerous ellipsoidal to sheet-like calcium-carbonate concretions were left behind and worked into a channel lag of rounded pebbles and cobbles (mean intermediate diameter of largest clasts from 5 to 35 cm), shaped by high-discharge streamflow along the active channel into longitudinal bars. Epiclastic granitic granule to pebble gravel and coarse sand derived from uplift cores to the west were deposited within the clast framework of these conglomerate bodies. Granitic sand

Figure 15. (A) Stratified fluvial volcaniclastics characterize the lower part of Upper Harrison valley fills. A channel within the Spoon Butte paleovalley filled with horizontally stratified and low-angle cross-stratified sandstone (cs) cuts flood-plain siltstones and sandstones (fp). Channel and flood-plain deposits are primarily fine-grained volcaniclastics, with only a small epiclastic fraction. Hammer is 30 cm in length. (B) Photomicrograph of very fine-grained, flood-plain, volcaniclastic sandstone comprising about 47 percent feldspar, 12 percent quartz, and 41 percent volcanic glass shards. Grains are more rounded and less well sorted relative to grains from volcaniclastic loess (Fig. 8C). (C) Rare lenses of epiclastic sandstone and gravel occur at and near the base of the fluvial volcaniclastic fill in the Spoon Butte paleovalley. Photomicrograph of trough-cross-bedded sandstone with grains of microcline and polycrystalline quartz indicates derivation from uplift cores to the west; composition is about 60 percent feldspar (3 to 4 percent microcline), 25 percent quartz (2 percent polycrystalline), 6 percent chert, mica-pyroxene-hornblende 4 percent. Note grain rounding and absence of volcanic glass shards.

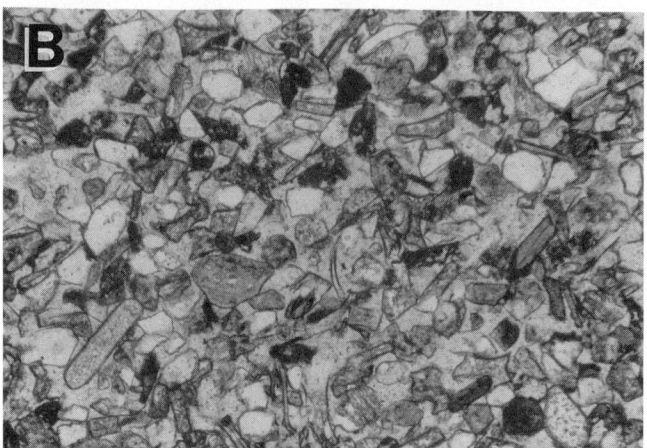

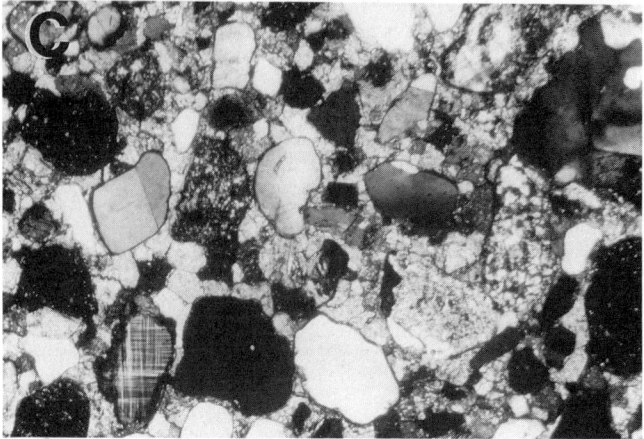

and granule gravel also occur as small, localized wedges of trough-cross-stratified sandstone deposited along valley margins (Fig. 15C).

Finest-grained sediments within the Spoon Butte paleovalley are represented by thin lenses of diatomite and sheet-like, tabular, freshwater limestones that record quiet-water deposition in shallow ponds on the valley flood plain. Scanning electron microscopy of diatomite reveals preservation of diatom frustules in association with opaline silica lepispheres (Fig. 16). Limestones are micritic, often infiltrated and replaced by silica, and similar in all respects to the limestones (lithofacies F2–LS) of the interchannel areas. Limestones associated with alluvial sediments are much less common than limestones within the interchannel reaches.

Most sediments filling the Spoon Butte paleovalley, however, are tuffaceous, fine- to very fine grained sandstones and siltstones, transported, deposited, and reworked in channels and over the flood plain. These sandstones and siltstones (Fig. 15B) differ from those of the Agate paleovalley in that they have a slightly higher percentage of epiclastic detritus (Table 4), and show a somewhat greater degree of grain rounding. The stratified sandstones and siltstones, and less prevalent limestones and tuffs, of the Spoon Butte paleovalley have been grouped into four subfacies (Skolnick, 1985): (a) very low angle to horizontally stratified sandstone, often deposited in association with scours containing lenses of poorly sorted lithic gravel and sand, succeeded vertically by trough- and occasional planar-cross-stratified sandstones; (b) vertically stacked sedimentation units, each made up of a horizontal, thin to medium bed of fine-grained sandstone, grading upward to siltstone and occasional thin diatomite; the sandstones display only small-scale, lower-flow-regime, sedimentary structures that include low-angle and trough cross-lamination, sometimes associated with horizontal and ripple lamination; (c) thin tabular sheets of micritic limestone; and (d) lenses of vitric tuff. Depositional environments indicated by these subfacies include (a) ephemeral or intermittent stream channel; (b) flood plain, (c) ephemeral flood-plain pond, (d) and waterlaid volcanic ash in flood-plain depressions.

At Spoon Butte the valley fill is dominated by (a) in-channel scour-and-fill sedimentation units, consisting of a concave-upward scoured base; a poorly sorted lag gravel; variably developed, thin- to thick-bedded, horizontally to very low angle stratified, fine-grained sandstones; and trough-cross-stratified sandstones that are incised and partly removed by superjacent

Figure 16. SEM photomicrograph of diatomite within fluvial volcaniclastic sediments of Spoon Butte paleovalley flood plain. Note opaline silica lepispheres (at arrows) adhering to the surface of diatom frustule fragments.

scour-and-fill units; and (b) laterally persistent, fining-upward sequences of sandstone-siltstone-diatomite, indicative of floodplain deposits. In (a), there is abundant evidence of the role of erosion and rapid sedimentation in the development of channel-fill sequences. In (b), vertical accretion without evidence of erosion between sedimentation units is the dominant theme in the flood-plain environment. The upper parts of sedimentation units are commonly pedogenic horizons overprinted by dense rhizolith-burrow complexes and heavily cemented by calcium carbonate.

Within the channel fills, the numerous concave-upward scours are of only limited lateral extent, but the channel floors are broad, flat, laterally extended surfaces. The general absence of lateral accretion deposits within the channels and predominance of vertical scour-and-fill deposits suggest vertical aggradation as the principal depositional mode within both channel and flood-plain environments, punctuated by periods of nondeposition in which pedogenic surfaces developed over varying time intervals. The channel-fill sequence common to modern ephemeral streams described by Picard and High (1973, p. 198), in which an erosional surface or scour is overlain by a lag deposit, inclined or discontinuous horizontal stratification, and trough cross-stratification, is readily observed within Spoon Butte valley channels. The alluvial architecture of the valley fill is a patchwork of (a) vertically aggraded, proximal scour-and-fill deposits, representing in-channel deposition; and (b) distal, thin- to medium-bedded, vertically accreted, sandstone-siltstone sequences, indicative of laterally equivalent flood-plain sedimentation.

In the Agate Monument paleovalley, the stratified volcaniclastics are largely well-sorted cross-stratified sandstones that contain almost no fossil mammals, whereas at Spoon Butte, the more poorly sorted and texturally varied ephemeral channel fills yield a good sampling of fossil vertebrates. The Agate Monument paleovalley has produced almost no mammal remains other than those associated with highly localized waterhole environments (see lithofacies F1–CM).

The nature and distribution of fossil mammal occurrences in these fluvial subfacies is controlled by local hydrodynamics and timing of streamflow, as well as the amount of sediment supplied during a particular sedimentation event. High-discharge flood events, with their accompanying large sediment volume, and rapid deposition in deep scours, are predisposed to preservation of vertebrate fossils encountered by the flood waters. Lower-flow-regime sheetflow over flood plains, in which thin sediment increments are transported and deposited across a broad valley floor, provides less opportunity for deep burial of skeletal material of moderate- to large-sized vertebrates, but nevertheless has resulted in burial *in situ* of subaerially weathered, partial skeletons on the flood plains of the Spoon Butte paleovalley. In contrast, well-sorted fluvial sandstones resulting from the temporally prolonged migration of bedforms in streams with low sediment supply are not likely to preserve fossil vertebrates in any abundance due to mechanical destruction of the fossils.

Fossil vertebrates found in the stratified volcaniclastics of the Spoon Butte paleovalley are primarily the disarticulated scattered remains of mammals, chiefly ungulates. Carnivores are present but rare. Most mammals occur as isolated bones, and as associated (not often articulated) elements representing an integrated portion of the skeleton (such as a partial limb or section of the vertebral column). Associated skeletal material is most common in the sandstone-siltstone units of the floodplain subfacies in which partial skeletons of camel, beardog, and entelodont have been discovered. Most bone, however, is found in the ephemeral channels where, as flood velocity declined, burial was rapid, the streams depositing significant sediment increments (as much as 2 m; 6.5 ft) at such times in broad local scours.

Bones are commonly weathered and scavenged prior to sediment burial. Evidence of bite marks and gnawing is present, and bone surfaces are often cracked and checked, some displaying very advanced stages of preburial subaerial weathering. Almost all bones have reached weathering stage 1 of Behrensmeyer (1978), and a number have attained stages 2 to 3. Abrasion of bone surfaces, however, is almost nonexistent, in strong contrast to the often strongly abraded skeletal elements found in the Agate Monument and Harper Quarry bonebeds. No major bone concentrations are found, fish are absent, and the remains of other aquatic vertebrates are nearly nonexistent.

Some mammal bone has been found in diatomite and in pond limestone within the Spoon Butte paleovalley flood plain. These vertebrate remains occur in the same manner as bones found in detrital flood-plain deposits. Associated partial limbs of large mammals as well as tiny bones of insectivores were collected from diatomite lenses. Diatoms and small planorbid gastropods were present in diatomite; aquatic pulmonate gastropods (lymnaeids) and ostracods are the only invertebrates found in the limestones.

Fossil wood is particularly rare in Arikaree sediments, but a small sample was collected from the surface of a sandstone outcrop within the Spoon Butte paleovalley. Although the wood was not found in place, there are no other rock units in the immediate area from which it could be derived by downslope movement, and it is likely that it comes from valley fill sediments. Thin sections of this wood reveal rings indicative of seasonal growth.

Interpretation. The stratified volcaniclastics are the episodically aggraded fill of wide, shallow, early Miocene valleys incised in the surface of the Hartville Table. Greatest stream energies and access to crystalline cores of the nearby Hartville uplift occurred early in the history of these valleys, demonstrated by the concentration of texturally coarse epiclastic debris in the lower part of the Spoon Butte paleovalley axis. The valleys then filled with sandy, fine-grained, tuffaceous sediments deposited by ephemeral or intermittent streams, involving both high-energy channelized flow and the accompanying shallow sheetfloods over adjacent flood plains. Such stratified volcaniclastics are restricted to the valleys and do not extend beyond the valley margins. The vast interchannel areas are devoid of any recognizable fluvial influences.

The Agate Monument paleovalley axis occurs only a short distance (24 km; 15 mi) northeast of the Spoon Butte paleovalley, and their axes are aligned along the same trend, suggesting that the two areas may be upstream and downstream segments of the same paleovalley. Unfortunately, the critical geographic area intervening between the two outcrop belts is covered by Quaternary dunes; however, the nature of the sediment infill in the two areas is in agreement with such an interpretation. Two superposed megafacies (stratified fluvial volcaniclastics beneath massive volcaniclastic loess) occur in the same spatial relation in both paleovalleys (Fig. 14). Fluvial sediments constitute the lower part of the fill of both the Agate Monument and Spoon Butte paleovalleys, and are overlain in each locality by 6 to 30 m (20 to 100 ft) of massive, volcaniclastic loessic sandstones. This volcaniclastic loess completes the filling of the valleys, resulting in abandonment of the earlier drainage pattern. Within both the Agate Monument and Spoon Butte paleovalleys, the upper loessic part of the valley fill includes silcrete paleosols that extend into and in some cases across the valleys. Above the silcrete, a prominent micritic limestone (Fig. 7, Agate Limestone) occurs in the Agate Monument paleovalley, indicating the presence of a shallow local topographic depression and lack of throughgoing streams at this time.

The more laterally confined and deeper Spoon Butte paleovalley, relative to the valley at Agate Monument, reflects, in its vertically aggraded scour-and-fill units with strongly erosive bases, a depositional environment dominated by periodic stream floods. Brief periods of high discharge were followed by rapidly waning stream energies and rapid deposition of fine-grained sediments. Upper-flow-regime deposits are more in evidence at Spoon Butte than at Agate Monument where they are extremely rare. Lateral to the flood-scoured channels, flood waters spread over the valley floor depositing thin beds of sand and silt. Water stood for brief intervals in shallow sheets and ponds, where diatoms bloomed, fueled by silica-rich floodwaters, and aquatic pulmonate gastropods and ostracods colonized ephemeral floodplain ponds in which fine carbonate mud accumulated. During long intervals between flood events, soils developed on the flat aggrading valley floor in company with the skeletal debris of resident mammals.

In contrast to the Spoon Butte valley floor, the Agate Monument paleovalley was broader and not as deeply incised into the surface of the tableland, and probably records the eastward shallowing of the Spoon Butte–Agate Monument paleovalley axis. The absence of deeply scoured and laterally restricted channels in the fill of the Agate Monument paleovalley, and the predominance of low sinuous-crested bedforms, forming extensive sand sheets across the width of the valley, indicates aggradation across a broad flat valley floor by widespread flood waters. The trough-cross-stratified sandstones produced by these sinuous-crested bedforms are lower-flow-regime deposits, as are the small-scale cross-laminated and rippled sandstones, and fine silt laminae, that compose the remainder of the Agate Monument paleovalley fill. Both of these sedimentation modes are consistent with the laterally extensive, nonerosive to weakly erosive, waning floodwaters that spread over a less confined valley floor.

Consequently, a regional model for Upper Harrison valley fill sediments deposited in the Spoon Butte–Agate Monument axial trend would incorporate deeper valley incision and confined flow in the more westerly reaches of the valley in Wyoming, and less confined, widespread, sheet-like flow with minimal erosion into bedload in the easterly segment of the valley at Agate Monument in Nebraska. Ephemeral, sandy braided-stream courses of low sinuosity and highly variable discharge were characteristic of the initial phase of valley filling. These streams possessed limited lateral channel mobility in their headward reaches, as demonstrated by a considerable volume of flood-plain sediments within the Spoon Butte paleovalley. Channel migration was influenced by the volume, energy, and duration of flood events, which were the intervals when most vertical scour-and-fill took place. This fluvial phase was followed by the second phase of valley filling when a blanket of volcaniclastic loess, punctuated by the development of silcrete duricrusts and shallow holomictic lakes, was deposited, creating a level regional plain of immense extent.

In the Agate Monument paleovalley, there was little difference in elevation between the valley floor and the nearly topographically equivalent interchannel reaches. Floods spread widely over the unconfined valley floor, joining thin fluvial sheet sands with interchannel loess deposits at the valley margins. During dry intervals, eolian processes winnowed the sandy bedload of these streams during times of reduced or nonexistent flow, and transported small amounts of mixed epiclastic and pyroclastic sediment into the interchannel environment. Massive or structureless sands within the fluvial valley sequences represent brief intervals of loess deposition on the valley floor that were not reworked by flood events.

The use of terms in the older literature such as "upland" and "lowland" for these rocks is of dubious merit, based on our recent recognition of the paleotopography of Upper Harrison environments. Low-gradient sandy-bedload streams traversing level plains of regional extent, only shallowly incised into a flat, gradually aggrading, regional surface, imply nearly uniform topography for these environments. The only topographic elevations anywhere adjacent to these plains are the buried uplift cores to the west, which at this time were nearly completely submerged under a regional blanket of Arikaree volcaniclastic loess, so that even the relief contributed by these features was minimal.

Pyroclastic air-fall events were the primary source of most of the fine-grained stream sediment, which was then further transported and reworked by fluvial and eolian processes. Wind undoubtedly carried volcaniclastic dust to both streams and interchannel plains. Degree of reworking was governed by the amount and seasonality of annual rainfall, distribution of vegetation, and the amount of sediment supplied to the system over given time intervals. The presence of prominent rhizolith and burrow concentrations at many horizons within the valley fills demonstrates that sedimentation events were episodic, in accord with the sporadic flow characteristic of modern ephemeral and intermittent streams, a point also supported by numerous mammal bones in advanced stages of subaerial weathering. Lack of significant grain rounding, especially in the Agate Monument paleovalley fill, considered together with the evidence for limited streamflow, must mean the amount of fluvial working of sediments in these valleys was minimal, and/or that rate of sediment supply was high enough to aggrade the fill without time for significant grain abrasion. Rainfall, and therefore stream flow, was probably seasonal (as suggested by growth rings observed in a single sample of fossil wood).

Textural and mineralogic similarity of these fine-grained volcaniclastics in both the valley fill and interchannel areas indicates a common origin from a distant volcanic source, and emphasizes the fact that these local environments and their biota would not have been preserved had it not been for the enormous volume of ash derived from more western source terrains during volcanic episodes in the early Miocene.

The manner of occurrence of mammal bone in these stratified volcaniclastics is the result of gradual accumulation and processing of skeletons derived from normal attritional deaths in and adjacent to the valleys over time. These are the scattered remains of mostly young, aged, and some prime individuals that died through the usual agencies of predation, disease, accident, and aging, and were scavenged and weathered in intravalley environments where rate of sedimentation was sufficient to bury a modest sample of their bones. The large percentage of ungulates relative to carnivores reflects normal mammalian predator/prey ratios. Here we have found no bone beds, nor even small concentrations belonging to multiple individuals of a single species; there is no evidence of fluvial agencies acting to concentrate large masses of vertebrate bone in particular sites, and it is evident that the bone sample that was buried within this valley is chiefly the skeletal residue that survived long intervals of subaerial weathering and/or happened at the time of death to be in temporal and spatial proximity to significant sedimentation events.

The range of sediment textures from diatomite and micrite to granule gravel and conglomerate records widely varying energies of deposition within the valley. Predictably, only a few durable bones occur in sediments indicative of sustained high fluvial energies deposited early in the history of these valleys. Subsequent intermittent streamflow aggraded the wide, level valley floor, reworking fine-grained bedload, and preserving the greater part of the mammalian skeletal material found in these deposits. Shallow ephemeral ponds on the flood plains accumulated diatomaceous and carbonate muds in which some bones were buried. Fossil soils at multiple levels in the stratified volcaniclastics indicate episodic sediment accumulation with sufficient time for processing (by animal activity and climate) of mammal bone subaerially exposed on the valley floor. Skeletons of most mammals that died within the confines of the valley were destroyed by weathering and scavengers during the intervals between sedimentation events. Eventually, fluvial activity ceased entirely, and the final phase of valley filling introduced fine-grained air-fall volcaniclastics that preserved vertebrate bone in the same manner as described earlier for the volcaniclastic loess lithofacies.

Waterholes: Intertongued pond marl and fluvial sandstone (Lithofacies F1-CM)

At the base of the formation, in erosional contact with subjacent Arikaree units, are occasional brilliant white lenses of altered calcareous tuff. These tuffs commonly contain a detrital fraction of very fine volcaniclastic sand and silt in a white siliceous glassy matrix. This matrix derives from the alteration of vitric shards, and comprises both authigenic montmorillonite and opaline silica. In addition to the detrital component, these altered tuffs contain micrite grains surrounded by siliceous cement or authigenic clay (Fig. 17), indicating that micrite was present during the initial phase of sedimentation in these shallow depressions. In addition, within some ostracods found in these tuffs, a small increment of micrite is found at the lowest point within the bivalved carapace where it has settled to form a geopetal fabric (Hunt, 1978, Fig. 15), also suggesting that micrite grains are a primary component of these deposits, and not a secondary calcite cement. Spar calcite, on the other hand, often acts as a void-filling cement in these tuffs. The incorporation of micrite, fine volcanic ash, and detrital silt and very fine sand in these deposits has resulted in their informal designation as pond marls.

These white lenses of calcareous tuff are distinguished from the hard, dense, cherty, lacustrine carbonate mudstones (lithofacies F2–LS) by their: (1) lack of chertification; (2) significant detrital grain content; (3) intertonguing relationship with stratified fluvial volcaniclastic sandstones; (4) lesser areal extent; and

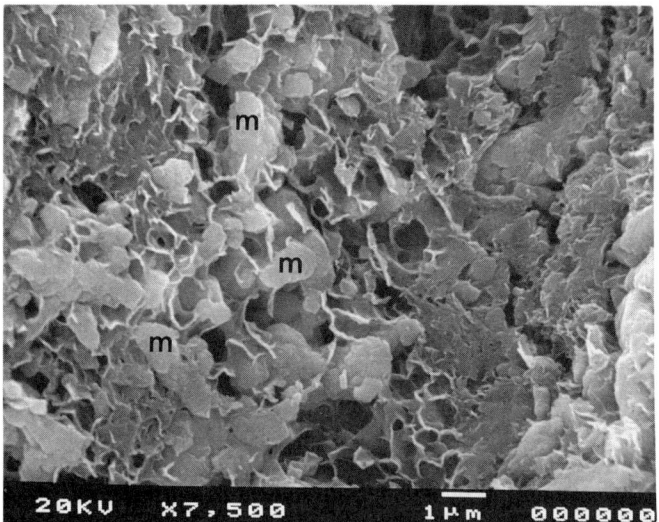

Figure 17. SEM photomicrograph of matrix of calcareous tuff (pond marl, lithofacies F1-CM) composed of micrite (m) and authigenic montmorillonite, forming a reticulate network between micrite grains. The source of montmorillonite appears to be dissolution of glass shards within the lithofacies.

TABLE 5. COMPARISON OF ACID (HC1)-INSOLUBLE RESIDUES FROM LACUSTRINE SILICEOUS LIMESTONES (LITHOFACIES F2-LS) AND FROM WATER HOLE POND MARLS (LITHOFACIES F1-CM), UPPER HARRISON BEDS, ARIKAREE GROUP*

Specimen Number	Lacustrine Limestones (%)	Pond Marls (%)
EC1	69.3	
EC2	68.1	
EC3	54.7	
EC4	65.1	
HQ2		43.2
HQ4		63.9
HQ5		51.3
SE1		80.2

*The insoluble residue of lacustrine limestones is almost entirely diagenetic silica that has replaced carbonate; the insoluble residue of pond marls is almost entirely detrital grains and glass shards.

(5) presence of major vertebrate bone beds. The dense, lacustrine carbonate mudstones are strongly silicified, do not contain any significant vertebrate remains, extend over considerable areas, lack detrital grains (except at their margins), and generally intertongue with massive volcaniclastic loess. Acid-insoluble residues of the altered micritic tuffs are entirely detrital grains, ash, and authigenic clay (with some opaline cement), whereas the residues of the dense lacustrine limestones comprise chiefly opaline silica that has infiltrated primary micrite fabric (Table 5 and Fig. 13).

Maximum thickness of these intertongued calcareous tuff/fluvial volcaniclastic sandstone sediment bodies is about 1.8 m (6 ft), with greatest diameter about 215 to 300 m (700 to 1,000 ft); hence, they are more restricted in areal extent relative to the lacustrine carbonates.

This lithofacies is best exposed at Agate Fossil Beds National Monument where it includes the sedimentation units incorporating the great mammalian bone bed (Fig. 6, F1–CM, waterhole). At Carnegie Hill, where the major fossil quarries were developed, the lithofacies comprises two end-members, white calcareous tuff and gray tuffaceous sandstone, that intertongue within the central part of the deposit. The gray sandstone is fluvial in origin, displaying ripple-lamination; small-scale trough and low-angle cross-lamination; and very small, shallow, scour-and-fill features. In places, the tongues of gray sandstone are cleanly separated from adjacent intertongued layers of calcareous tuff. Boundaries between the two lithologies are sharp and distinct. But over much of the central part of the deposit, sharp boundaries between the two lithologies can be traced laterally into highly contorted bedding, in which the white calcareous tuff and gray sandstone tongues are disrupted, distorted, and mixed together. Within these contorted sediments, we have found sand-filled track molds of large mammals such as entelodont and camel. At the northwestern margin of the lithofacies at Carnegie Hill, the gray tuffaceous sandstone incorporates tongues of fluvial lithic granule to pebble gravel that indicate local coarsening of detrital sediments toward the west.

At Agate Monument and Niobrara Canyon (Figs. 4 and 7), the two localitites where the lithofacies has been identified, its base lies on a topographically irregular erosion surface representing the scoured floors of wide, shallow, abandoned stream channels. At Carnegie Hill, for example, the channel has a minimum width of 85 m (280 ft), and is cut into underlying Arikaree rocks to a maximum depth of 1.4 m (4.5 ft). A transverse cross section of the Carnegie Hill channel is asymmetric, the northern margin rising abruptly to form a low vertical bank a few feet in height, whereas at the southern margin the profile rises gradually without relief. Angular to subrounded blocks derived from the channel banks and floor, and from indurated silcrete paleosols incised by these channels, occur as large isolated clasts within the base of the lithofacies. The silcrete clasts are invariably derived from the terminal paleosol of the subjacent Harrison Formation, which developed as a regional geomorphic surface prior to the incision of the first Upper Harrison channels in the region.

Buried within the fine-grained sediment filling these abandoned channels are dense accumulations of mammal bones, so heavily concentrated in some places that they may be termed "bone beds" (Fig. 18). In the entire formation, only in this lithofacies have such bone concentrations been found. The two principal bone beds occur at Harper Quarry (Hunt, 1978) and the Carnegie and University Hill Quarries at Agate Fossil Beds National Monument (Peterson, 1909; Matthew, 1923). A minor and

Figure 18. Mammalian bone bed at Carnegie Hill, Agate Fossil Beds National Monument, Nebraska, consisting of bones of the small rhinoceros *Menoceras arikarense* and the chalicothere *Moropus elatus*. Bones occur in intertongued calcareous tuff–fluvial sandstone (lithofacies F1–CM) deposited in an early Miocene waterhole that filled with fine-grained ash-rich sediment.

doned channel segments. The presence of fluvial sands and lithic gravel in these waterholes demonstrates their proximity to active channel environments transporting and reworking fine-grained sandy bedload.

These intertongued calcareous tuffs and fluvial sandstones record a transitional interval in the geomorphic history of the region when valley incision had ended and aggradation of the valleys had just begun. The geometry of several of these waterholes and their sedimentary fill is particularly well preserved; the Agate Monument and Harper Quarry bone beds are particularly good examples. Imprinted in calcareous tuff of the Carnegie Hill bone bed at Agate Monument are track molds of large ungulates, demonstrating that mammals bioturbated these sediments, thereby contributing to the deformed and disturbed bedding common to the lithofacies. A ponded setting is also corroborated by the presence of ostracods, pulmonate gastropods, charophytes, and diatoms.

That these waterholes were ephemeral is suggested by the absence of aquatic vertebrates commonly associated with permanent water. Invertebrates and plants in the pond marls belong to groups that typically colonize rigorous transient environments such as ephemeral ponds where seasonal desiccation and shallow water permit only a hardy biota to survive (Hunt, 1978). The nonmammalian biota present in this lithofacies rapidly repopulates such ponds today, and are often among the first migrants to arrive in such ephemeral aquatic environments.

The diatom assemblage from one of these deposits (Harper Quarry bone bed) has been studied, and is typical of diatom floras found today in warm, shallow, eutrophic, alkaline ponds (D. G. Maroney, unpublished data).

TAPHONOMY OF MAMMALIAN BONE BEDS

Taphonomic analysis provides useful evidence concerning the origin of bone beds associated with lithofacies F1–CM (see Table 7). An earlier study of the Harper Quarry bone bed (Hunt, 1978) in Niobrara Canyon was continued in the summers of 1985–1986 by excavation and analysis of the Agate Monument bone bed. These bone beds are attributed to drought-induced death events that, together with lithofacies analysis, indicate a seasonally dry climate within a semiarid continental interior. Furthermore, the moderate alkalinity of these waterholes, the presence of calcium carbonate in the form of micrite, and their probable rapid filling by volcanic ash, provided an ideal environment for preservation of bone.

Excavations by the University of Nebraska in 1986 investigated the nature of the Agate Monument bone bed at Carnegie Hill, the site of the principal fossil quarries for which Agate Monument is world famous. We discovered that the intertongued pond marl–fluvial sandstone lithofacies, which is about 1.5 m (5 ft) in maximum thickness, can be subdivided into four sedimentation units (A through D) that, in turn, can be grouped into two sedimentation events (A–B, C–D, Fig. 19). The two sedimentation events are well defined within the central part of the channel

less well known bone bed called the American Museum–Cook Quarry (Hunt, 1972) is found a few miles north of the Carnegie Hill bone bed on the northern margin of the Agate Monument paleovalley.

In addition to mammals, aquatic invertebrates and plants also occur in this lithofacies, primarily in the calcareous tuffs, and include freshwater ostracods, pulmonate gastropods, charophyte algae, and diatoms. As in the silicified lacustrine limestones discussed earlier, there are no fish or other aquatic vertebrates.

Interpretation. The association of calcareous tuffs with ripple-laminated, fluvial, volcaniclastic sandstones in topographically low areas along the axial trend of these paleovalleys, and the absence of such deposits in interchannel environments indicate that this lithofacies accumulated in standing or ponded water situated along the drainages. These waterhole deposits are particularly prevalent during the early history of valley filling in the region, and most appear to be ponds that developed within aban-

axis where each event comprises a marl (calcareous tuff laterally grading to fine sandstone) overlain by a fluvial sandstone tongue. The channel axis (N9-N34, Fig. 19) is the deepest part of the waterhole where the thickest fill was deposited: within the axis, the basal marl unit containing the lower bone bed is 30 to 45 cm (1 to 1.5 ft) thick. Above it are 5 to 45 cm (a few inches to 1.5 ft) of finely ripple-laminated, gray, fluvial sand, overlain by as much as 23 cm (9 in) of a second marl unit. This second marl grades upward into 60 cm (2 ft) of gray, rippled, fluvial sandstone, which is then overprinted by a paleosol with a well-developed petrocalcic horizon, indicating a long period of nondeposition. To the southeast, all sedimentation units grade laterally to pond marl (calcareous tuff). To the northwest, all sedimentation units grade laterally into tuffaceous fluvial sandstone, and at the west bank of the channel the lower two sedimentation units (A, B), composing the first sedimentation event, end against the channel margin (between N34 and N40, Fig. 19). Only the upper sedimentation units (C, D) continue to the northwest, overtopping the low channel bank.

The principal or lower bone bed at Carnegie Hill is contained within the lowest sedimentation event, in fact within sedimentation unit A. Unit A is overlain by the unit B sandstone, which contains no bones. Resting on the surface of unit B and buried within unit C is an upper bone bed, in which bones are less numerous than in the lower bone bed. The upper bone bed within unit C is overlain by the terminal sedimentation unit D, a fluvial sandstone, which in turn is overprinted by a paleosol horizon indicating a pause in fluvial deposition within the channel. In the subsequent discussion, unless otherwise indicated, the term "Agate bone bed" refers to the lower bone bed within sedimentation unit A.

The Agate bone bed is primarily made up of the skeletal remains of only two mammals: the small rhinoceros *Menoceras arikarense*, and the large chalicothere *Moropus elatus*. In the deposit, rhinoceroses number in the hundreds, but no more than 50 to 75 chalicotheres are present. A third mammal, the entelodont *Dinohyus hollandi*, is represented by two partial skeletons, and by widely dispersed bones of a few other individuals. Also scattered through the bone bed are isolated, highly abraded bone fragments and teeth of horse, camel, oreodont, and carnivore; however, skeletal remains belonging to mammals other than the small rhinoceros, chalicothere, and entelodont are extremely rare. Greater species diversity was found in the Harper Quarry bone bed (Hunt, 1978), where a minimum of 12 kinds of mammals were discovered in a quarry sample of 545 bones, and no single species dominated the bone assemblage.

Mapping of fossil distribution in the Agate bone bed indicates that both thickness of the bone bed as well as bone density over a given area vary greatly. The first excavators in these quarries, who had ample opportunity to observe bone-bed dimensions, recorded a maximum thickness of 45 to 60 cm (1.5 to nearly 2 ft). The thickest portion of the bone bed, with dense packing of many elements, occurs in the topographically lowest part of the waterhole, referred to as the center of the deposit in the subsequent discussion, although it may be somewhat eccentrically situated if the true geometry of the waterhole could be fully reconstructed. In peripheral areas the bone layer is thinner, with greater dispersion of elements, often amounting to only the thickness of a single scattered layer of bones, and some places are devoid of bones. At Harper Quarry, the bones occur as a single layer 15 to 25 cm (6 to 8 in) thick that varies in elevation about 55 cm (1.8 ft) from the base of a shallow channel scour to its margin. Some areas of the bone bed display dense bone concentrations, whereas adjacent areas are barren of bones (Hunt, 1978).

Within the central part of the Agate bone bed, bone density commonly exceeds 100 bones per square meter, and in peripheral areas averages about 40 bones per square meter (excluding barren areas of the quarry floors). At Harper Quarry, the maximum bone density is 40 to 50 bones per square meter, comparable to the peripheral areas of the Agate bone bed, and there are large areas of the bone bed that average about 10 bones per square meter (this does not include numerous small bone fragments with a long axis less than 2 cm that are present throughout the quarry).

Time of subaerial exposure to the environment between death and sediment burial can be estimated by the degree of disarticulation and weathering of bones (Hill, 1979; Hill and Behrensmeyer, 1984; Behrensmeyer, 1978). Nearly all bones in the Agate bone bed are disarticulated (Fig. 18). Yet some partially articulated bones have been observed; they occur in the central part of the deposit where the denser bone concentrations were found. These thick portions of the bone bed naturally were exploited by the early field collectors, who cut out large sections or "blocks" for removal to their respective institutions. Many blocks taken from the quarry floor in the Southwest Excavation at Carnegie Hill show some residual articulations (Fig. 20), but ligaments and other connective tissues binding these skeletons have usually decayed to the extent that most bones have separated from each other, and been slightly displaced. Matthew (1923) noted that in his experience in working with the Agate rhinoceros skeletons the bones of single individuals could be found together within a small area. Whereas this may be true near the center of the bone bed where bone concentration is highest, it is not the case at the periphery of the deposit where bones are scattered and almost no articulations remain.

One of the rare skeletons for which partial articulation was confirmed in the bone bed was the Carnegie Museum entelodont *Dinohyus*, found at the edge of the Southwest Excavation on Carnegie Hill (Peterson, 1909). Its skull, jaws, and vertebral column were nearly intact; ribs, sternum, girdles, and limbs were nearby, in association but scattered. Such partially articulated individuals are unknown in the Harper Quarry bone bed where no articulated material has been found.

Subaerially weathered bone is not common anywhere in the Agate bone bed. Where it occurs, weathering is limited to development of fine cracks in the surface of compact bone. In a sample of 684 bones derived from 11 m^2 of the Agate bone bed, 94 percent could be assigned to Behrensmeyer's (1978) weathering

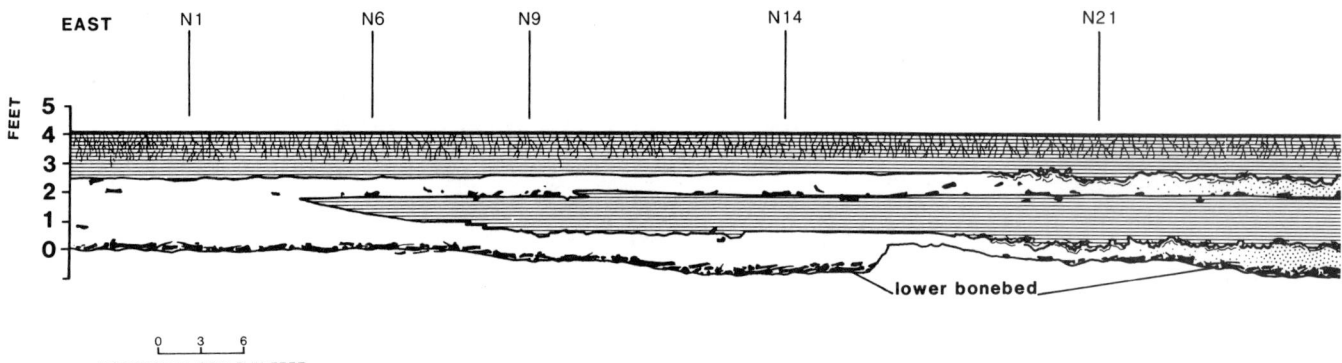

Figure 19. Stratigraphic cross section of the Agate Monument waterhole, showing the sedimentation units (A through D) enclosing the lower and upper Agate bone beds, North Excavation, Carnegie Hill, Agate Fossil Beds National Monument, Nebraska. N1 through N49 are locations of measured sections used in construction of the profile.

stages 0 or 1. Skeletons from the central part of the bone bed show little evidence of weathering, and in some cases, bone surfaces are extremely fresh, also displaying no evidence of surface abrasion. Such bones have undergone little postmortem processing of any sort. On the other hand, at the periphery of the bed, some bones exhibit fine linear cracking and occasional flaking of the surface, indicating at least a short interval of subaerial exposure. Immersion in water or moist sediment of approximately neutral pH retards bone weathering; this probably explains the unweathered condition of many skeletons near the center of the Agate bone bed.

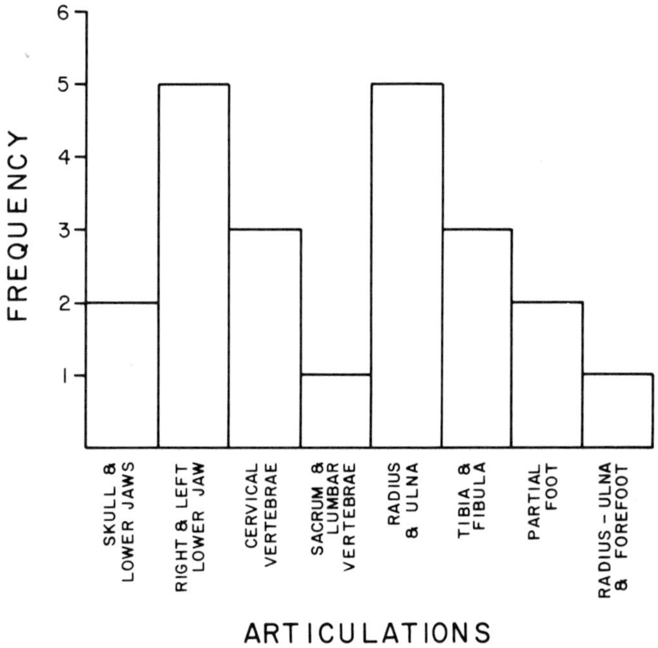

Frequently, prolonged subaerial weathering of the skull and jaws results in loss of teeth from the sockets, and separation of the skull into segments along loosened sutures, especially in younger animals. This is not apparent in the central bone bed, where most skulls and jaws retain the full dentition and are usually intact, but in peripheral areas, fragmented skulls and jawbones lacking teeth are often found.

In the Harper Quarry bone bed, the range of weathering stages is similar to that at Agate. Bones in Behrensmeyer's stages 0 to 2 are present at both sites, but at Harper Quarry there is a higher percentage of moderately weathered material: almost all bone surfaces can be placed in stage 1 or 2, and there is a greater proportion of bones at stage 2 than at Agate. At Harper Quarry, only fragmented skulls have been found in the bone bed. In different parts of the bone bed, bone preservation ranges from fragmentary jaws, isolated teeth, and broken bone fragments to complete lower jaws of both ungulates and carnivores with full dentitions, intact limb bones, ribs, and vertebrae.

In the topographically lowest part of the Agate waterhole where the bone bed is thickest, the bones occur in a variety of spatial orientations, often in unstable positions. At the periphery of the waterhole, where the bone bed is thin, the bones are scattered as a monolayer, and tend to lie flat on the surface of the bed in stable orientations (Fig. 21). Associated with the parts of the bone bed where bones are in unstable alignments are tracks of large mammals that penetrate and bioturbate the surrounding sediment. At Harper Quarry, mammal tracks, even if present, could not be easily identified in the homogeneous pond marl; however, very few bones in the quarry show steeply inclined long axes—85 percent of bones with a designated long axis plunge less than 20 degrees, and 92 percent plunge less than 30 degrees (Hunt, 1978).

Proximity of skeletons belonging to a single species is unique to the Agate waterhole and does not occur at Harper Quarry. About 20 chalicotheres were found together, some skeletons

Figure 20. Frequency of residual skeletal articulations for the rhinoceros *Menoceras arikarense*, counted in a rectangular slab 2.2 m^2 (24 ft^2) in area removed from the Agate bone bed, Southwest Excavation, Carnegie Hill, and now in the Denver Museum of Natural History. These remnant articulations are among the last to dissociate in disarticulation studies of living ungulates about the size of the Agate *Menoceras* in East African savanna environments.

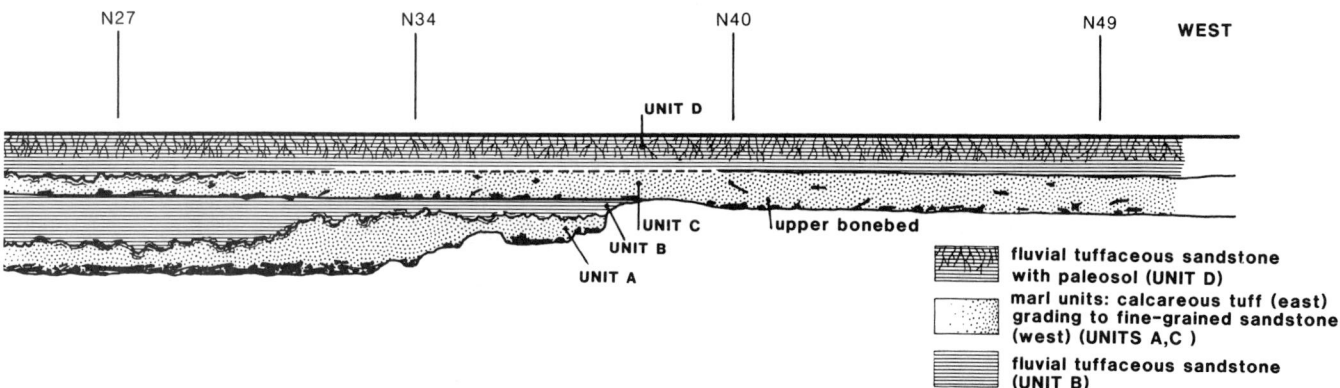

nearly complete (Fig. 22), segregated in a topographically low area along the west margin of the waterhole (American Museum Chalicothere Quarry, Southwest Excavation, Carnegie Hill). Skeletons of this chalicothere group have been partly mixed together yet have been only slightly displaced within the local area where their carcasses are concentrated. Other areas in and near the deepest part of the waterhole contain masses of densely packed rhinoceros bones. Elsewhere, both near the center and at the periphery of the deposit, one finds scattered bones of rhinoceros and chalicothere mixed together, including an occasional entelodont bone. In contrast, at Harper Quarry the bones of all species are randomly mixed over the quarry floor.

Because of differences in size, shape, and density, bones of a skeleton are not equally susceptible to sorting by a current of flowing water. If strong currents influenced the bone bed after the skeletons were disarticulated, we would predict that many of the lighter elements, and those readily lifted into a moving current by virtue of their shape, would be removed. Since mammalian bone beds associated with this lithofacies intertongue with fluvial sediments, we wish to know if the skeletons were winnowed by streamflow, and bones transported prior to burial.

Chalicothere and rhinoceros skeletons in the deeper part of the waterhole often show a high percentage representation of bones of the skeleton (Fig. 22, Southwest Excavation). Most of these skeletons were fully disarticulated, yet the bones of a single individual remained together, including light, readily transportable elements that would be the first to be moved by stream currents. On the other hand, in the peripheral areas of the waterhole, associated skeletons are nonexistent, and the bone bed is essentially a single layer of scattered bones: degree of skeletal

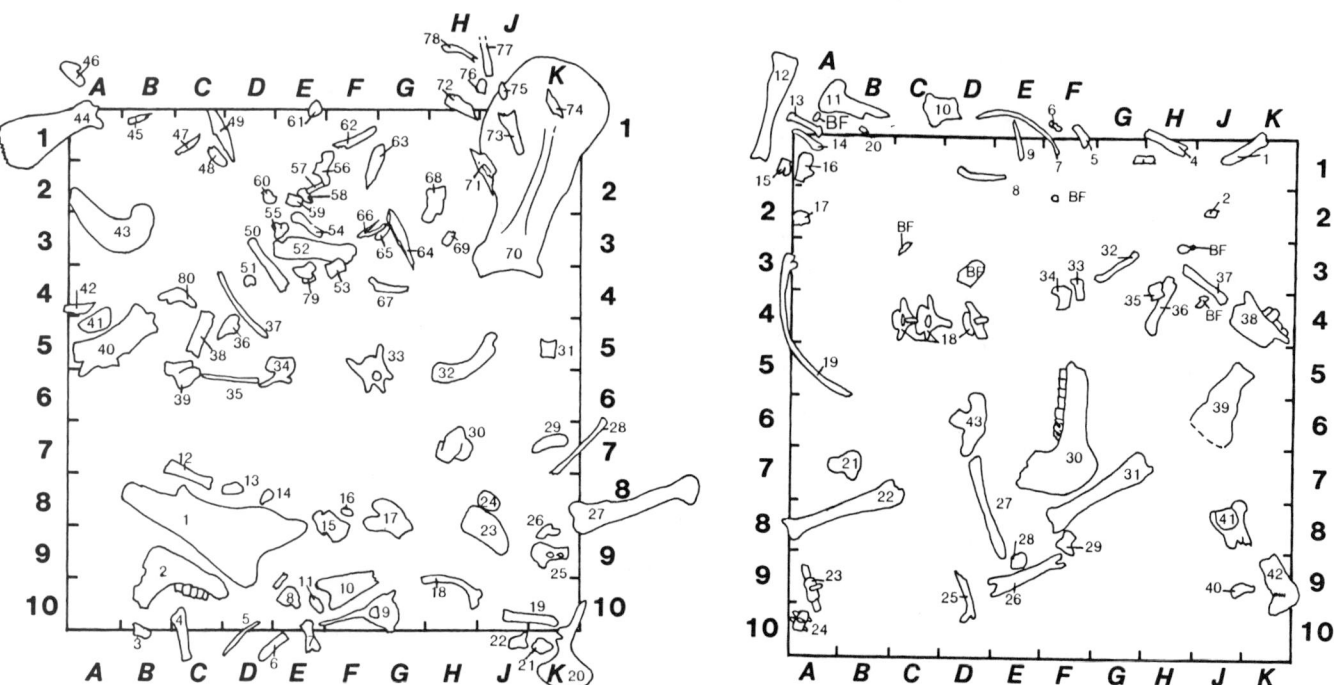

Figure 21. Plan view of two representative 1-m² areas of the lower bone bed, South Excavation, Carnegie Hill, developed by the University of Nebraska in 1986 at the periphery of the Agate waterhole, where the lower bone bed is a thin monolayer of disarticulated and scavenged bones of rhinoceros and chalicothere that lie in stable orientations on the floor of the old channel.

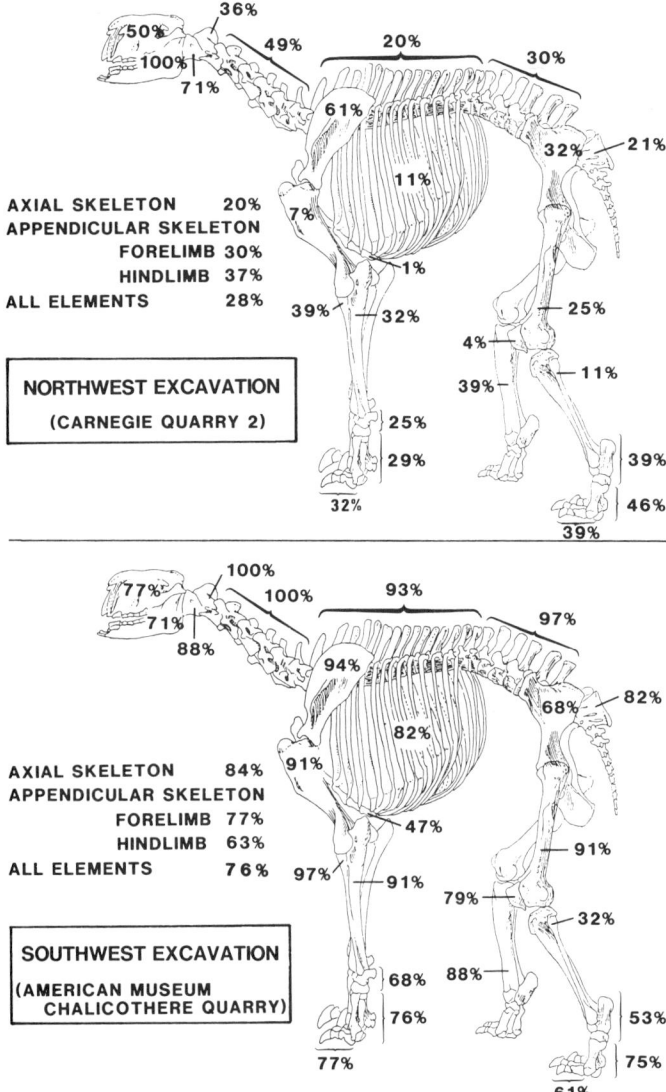

Figure 22. Estimates of skeletal completeness for the chalicothere *Moropus elatus* found in the Agate bone bed, based on determination of minimum number of individuals. Top: Northwest Excavation samples the upper Agate bone bed and the periphery of the waterhole, and shows low percentage values for skeletal completeness, with an average for all skeletal elements of 28 percent. MNI=14. Bottom: Southwest Excavation samples the lower Agate bone bed and the more central part of the waterhole, and has high percentage values for completeness, with an average for all elements of 76 percent. MNI = 17. Bones in the Northwest Excavation are dispersed, scavenged, with crania badly fragmented; bones in the Southwest Excavation in many cases still are associated as individual skeletons, have undergone very little processing by scavengers, and most crania are intact.

completeness is difficult to estimate, but where such estimates are possible (Fig. 22, Northwest Excavation), they are lower than near the more central deeper part of the waterhole. However, sorting by flowing water is probably not the cause of the pattern of bone distribution observed in the more peripheral parts of the deposit, since the bones that remain display a wide range in size and weight, from large, heavy limb bones that would move only in the strongest currents to small, light bones easily transported under most lower-flow-regime conditions.

Additional insight into the question of stream transport of the Agate bone bed skeletons can be gained by considering the relative size of individuals making up the small group of chalicotheres that lies together along the western margin of the waterhole (Fig. 22, Southwest Excavation). As noted earlier, the bones of a number of these chalicotheres are only slightly displaced, making it possible to identify individual skeletons (Matthew, 1923; Coombs, 1975). Study of the degree of epiphyseal fusion in these skeletons, coupled with stage of tooth eruption and wear, permits determination of ontogenetic ages of these animals. Small juveniles, young adults of intermediate sizes, older females, and large adult males are present. If stream transport of these animals occurred, they were moved as intact carcasses since the bones of individual skeletons are in proximity in the bone bed. It appears improbable that streamflow would maintain the association of many individuals of a single species during transport of carcasses because of the great range in body size within this group of animals, even if transported as dried, intact, hide-encased carcasses (such transport of hide-bound dehydrated ungulate carcasses is documented in East Africa in modern semiarid environments). Because of the great size of these chalicotheres (large males would weigh 1000 kg), the transporting agency would have to have been a major flood event, and small and large animals would have to have been transported together, as a group, without regard for differences in weight and size. In addition, the waning flood would have to deposit the carcasses in a particular part of the waterhole. Not only is such segregation of carcasses of a single species hydraulically unlikely, but there is no evidence in the associated sediments for a major flood event. In fact, the segregation of carcasses within the waterhole is in better agreement with an *in situ* death event. The same argument can be applied to the rhinoceroses in the bone bed (based on evidence of segregation of small groups of rhinoceros), and leads to the conclusion that these animals underwent little or no fluvial transport as carcasses. It is more likely that they walked into the waterhole and died at or near the places where their bones are found.

Following death and disarticulation of rhinoceros and chalicothere in the Agate bone bed, the bones were either never strongly aligned by flowing water or, if such alignments existed, they were later erased by subsequent animal activity in the bone bed. In fact, the thicker parts of the bone mass display an interlocking bone fabric that probably prevented alignment by low-energy currents. Rose diagrams depicting the axial orientations of long bones from the Carnegie Hill bone bed (Fig. 23) were constructed from data collected during original field excavations in 1986 and from the large blocks removed by the earlier excavators. Orientations were measured for bone samples from both the calcareous tuff subfacies and fine-grained sandstone subfacies of the bone-bed marls (units A and C, Fig. 19). Bones buried in calcareous tuff invariably display a pattern that cannot be distin-

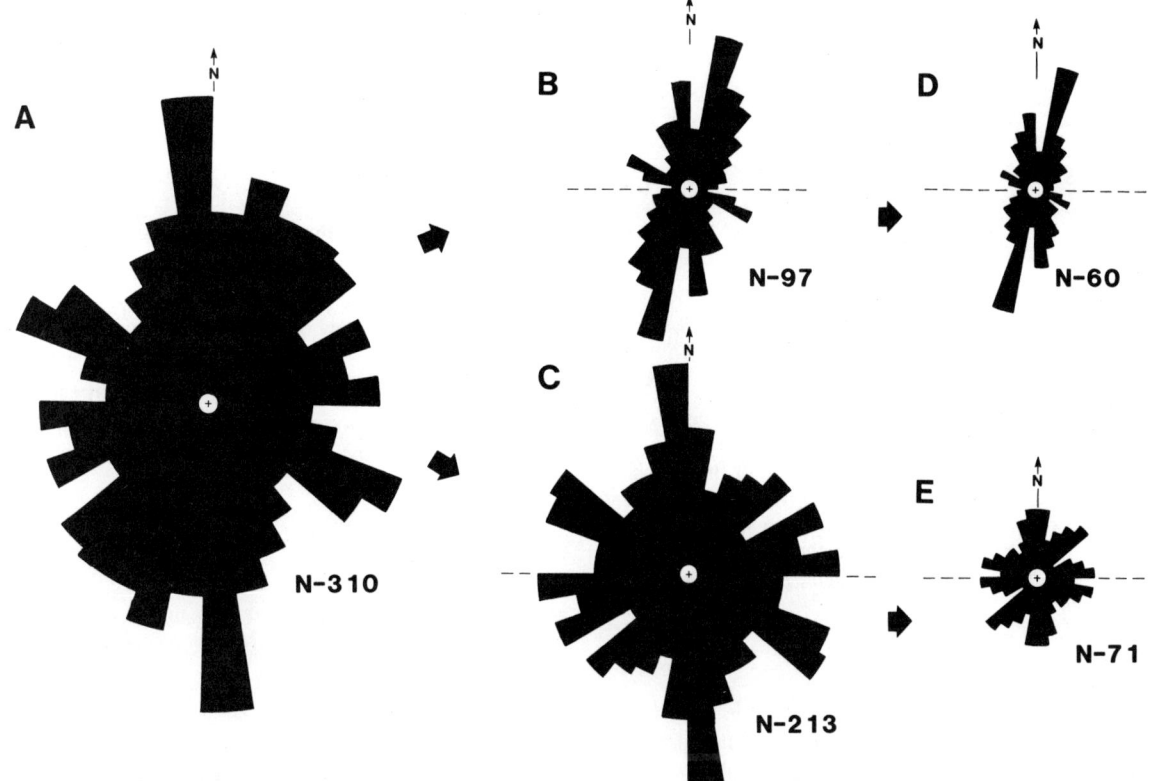

Figure 23. Mirror-image rose diagrams of orientations (azimuth) of the long axes of elongate bones from the lower Agate bone bed, based on 1986 University of Nebraska excavations of 8 m^2 of quarry floor in the North, Southwest, and South Excavations at Carnegie Hill. A, all 310 bones from both calcareous tuff and tuffaceous sandstone (meters 1 through 8); B, subset of 97 bones from tuffaceous sandstone only (meters 1 and 2); C, subset of 213 bones from calcareous tuff only (meters 3 through 8). D, subset of 60 bones from tuffaceous sandstone (meter 1 only); E, subset of 71 bones from calcareous tuff only (meter 8 only). A = B + C; D is a subset of B, and E is a subset of C. Measurements are grouped in 10-degree classes; north is at top of figure.

guished from a random distribution (Fig. 23C and E), but measurements of bones buried in fine-grained sandstone suggest preferred alignment for some samples (Fig. 23B and D). This is interpreted as a residual local alignment of some parts of the bone bed influenced by stream currents early in its history, and not as alignment produced by deposition of the ripple-laminated fluvial sandstone that laterally grades into carbonate mudstone. Stream currents having sufficient energy to align bones from animals the size of rhinoceros and chalicothere would have removed carbonate mud and fine sand from the channel axis, and we see no evidence of this. Once the waterhole began to fill with sediment, this process was not interrupted by any significant erosional events.

The Harper Quarry bone bed also produced a polarity rose that suggests a pattern of preferred bone alignments developed early in the history of the deposit under a fluvial regime (Hunt, 1978, Fig. 11). Thus, in both the Agate and Harper Quarry sites, if preferred bone alignment occurred, it was produced prior to the infilling of these waterholes with fine sediment.

Despite the lack of evidence for sorting of skeletons by energetic stream currents, many bones in the Agate bone bed show moderate to strong surface abrasion. There is also a correlation between the amount of abrasion and the degree of completeness of most skeletons. The least abraded bones often belong to the more complete and less disturbed skeletons, usually from the deepest part of the waterhole. Highly abraded bone is often found at the margins of the deposit where it represents the general condition of the scattered unassociated skeletal debris (Fig. 24). Bones mixed with the thin sandy lag gravel that occurs locally at the base of the deposit are particularly well abraded. Early excavators in the bone bed reported that strongly abraded bones often occurred at the base of the bone layer underneath a thick mass of generally unabraded skeletons (Peterson, 1923).

Most of this abrasion probably occurred within the confines of the waterhole, based on lack of evidence of hydraulic sorting of skeletons, and also because of the presence of numerous selectively abraded bones (especially astragali) in which only one face exhibits significant wear while the opposite side is nearly unworn. Such bones must have been in only limited contact with an abrading medium, and could not have been repeatedly tumbled or rolled along a bed surface. This implies an oscillating motion of the bone while held in a constant orientation relative to the

Figure 24. Scavenged and abraded bones are typical of the peripheral areas of the Agate bone bed. Note ends removed from long bones, and green bone fractures of large bone fragments to left of scale (upper edge in cm). Abrasion follows scavenging since fracture edges are smoothed. This processed bone is much more common in the bone bed than earlier excavators had reported.

abrading agency, presumably the bed surface. Movement of water and sediment within the waterhole were not energetic enough to tumble or transport most bones, but must have been sufficient to oscillate bones resting on the channel floor, resulting in the observed patterns of selective abrasion. Very likely the weight and interlocked fabric of the thick bone mass acted to keep bones at its base in place so that stream currents flowing through the deposit could not transport such bones, but would simply oscillate them on the floor of the channel. Also, these same bones were probably further abraded by sediment/water set into motion by the feet of large numbers of milling animals, coming to the waterhole to drink (similar bone abrasion in a small South African waterhole frequented by domestic goats has been described by Brain, 1981). Hence, most heavily abraded bones probably were produced at the base of the bone bed where bones were held in direct contact with the sandy substrate, and at the margins of the deposit where the daily activity of animals using the waterhole continually agitated the water and sediment.

In the Harper Quarry bone bed, the number of significantly abraded bones is high (Hunt, 1978, p. 25), and a pattern of selective abrasion like that seen at Agate is not apparent. Probably most abrasion was produced by streamflow before circulation through the abandoned channel segment was curtailed, prior to deposition of the sediment plug that fills the waterhole.

Throughout the Agate bone bed, the results of animal activity in modifying bone are everywhere evident. Animal-processed bone occurs in almost all parts of the deposit, but is most common in marginal areas, where scavenged bones make up the major part of the bone assemblage (Table 6). We found abundant evidence of bone processing at the periphery of the bone bed in our 1986 excavations (Fig. 24), and have also observed that some of the large blocks of fossils, taken from the deeper part of the waterhole where the bone bed is notably thick and dense, are also made up of highly processed bone (Fig. 18). Yet there is no doubt that some central areas of the waterhole contained masses of skeletons without any bone processing in evidence (Table 6, meter 6), and these areas also show little or no abrasion or sorting.

Animal activity is manifested in these waterhole deposits as: (1) bone modification through scavenging, and (2) bone modification by trampling. Both processes can be documented in the Agate bone bed.

Biting and gnawing on bones is demonstrated by percussion fractures of the bone surfaces (circular to elliptical in form), as excavated (gnawed) zones located primarily at the ends of long bones, and as green bone fractures (Fig. 24). In modern environments, this kind of bone modification can be directly attributed to mammalian predators and scavengers (Haynes, 1980, 1982; Binford, 1981). In the Upper Harrison fauna, large carnivorans and entelodonts fill this role. The large body size and robust dentitions of four Upper Harrison carnivorans (three amphicyonid beardogs and a large wolverine-like mustelid) suggest all would have been capable of bone processing of the type seen in the Agate and Harper Quarry bone beds.

During excavation of the Agate bone bed in 1986, we discovered large percussion fractures in the shaft of a chalicothere humerus (Fig. 25). The smallest, a nearly circular flat-bottomed depression, is 2 cm in diameter, and adjacent ones are much larger. The maximum diameter of worn canine teeth of old indi-

TABLE 6. PERCENTAGE OF MAMMAL BONES BROKEN PRIOR TO SEDIMENT BURIAL IN THE AGATE BONEBED (LITHOFACIES F1-CM), CARNEGIE HILL QUARRIES, AGATE FOSSIL BEDS NATIONAL MONUMENT, NEBRASKA*

Test Meter	Preburial Breakage	
	Yes	No
1	51	49
2	38	62
3	57	43
4	53	47
5	62	38
6	9	91
7	71	29
8	73	27
9	65	35
10	74	26
11	86	14

*Data from 1986 excavations by the University of Nebraska. All bones excavated from eleven one-meter square areas were examined for preburial breakage. Meters 3, 4, 5, 9, 10, 11 sample the periphery of the waterhole; meters 6 and 7 sample the thickest part of the bonebed within the topographically lowest area of the waterhole; meters 1, 2, 8 sample an area of intermediate depth between the periphery and more central part of the waterhole.

Figure 25. (A) Entelodont artiodactyl track pressed into calcareous tuff and later filled by gray fluvial sand, observed in sedimentation unit A (Fig. 19) containing the lower Agate bone bed, Southwest Excavation, Carnegie Hill. Note bones on bone bed floor immediately below footprint. Trampling of the waterhole mud is the cause of contorted bedding seen in the deeper part of the fill, and is probably also responsible for the unstable orientations of many bones. (B) Large bite marks in the shaft of a chalicothere humerus, South Excavation, Agate bone bed. Size and geometry of these percussion fractures suggest they were made by scavenging entelodonts.

viduals of the largest beardog living at that time attains only 1.2 cm (canines in old individuals develop flattened tips; in young beardogs, the canine is sharp and would not make a flat-bottomed percussion fracture). In fact, among all species of the Upper Harrison mammal fauna, only flat-topped premolars of the huge entelodont *Dinohyus* fit these percussion fractures, indicating the probable scavenging role played by entelodonts.

Much of the observed bone breakage can be attributed to green bone fracturing, so scavenging occurred while bone was still relatively fresh. Most scavenging probably took place while at least some connective tissue remained on the bones. In almost all instances of scavenged bone in the bone bed, abrasion overprints the broken surfaces, demonstrating that abrasion occurred after bone processing was completed (Fig. 24). The Harper Quarry bone bed also consists of similarly scavenged skeletal remains, many bones showing bite marks, gnawing, and preburial breakage attributable to green bone fractures.

Trampling within the Agate bone bed is demonstrated by the discovery in 1986 of large ungulate tracks pressed into the limy tuffaceous mud of the waterhole and filled with fluvial, gray, fine sand. Some can be identified as those of entelodont on the basis of their cloven-hoofed (artiodactyl) geometry and large size (Fig. 25), and others as the tracks of camels. Track bioturbation is most easily recognized where layers of gray fluvial sand and white calcareous tuff are intertongued. In such areas, the contrast between track molds pressed into the white mud and their gray sandy infill is particularly striking. Contorted bedding in the central part of the waterhole indicates trampling occurred there. But in other areas, particularly the periphery of the waterhole where only a single lithology is present, bioturbation is difficult to identify. Importantly, the bioturbated areas of the bone bed are also the areas where unstable bone orientations are observed to be particularly common (Fig. 26); hence, the trampling of waterhole muds is probably responsible for reorientation of bones within these parts of the bed.

The foregoing evidence of disarticulation, weathering, abrasion, and animal modification of bone in the Agate and Harper Quarry bone beds (Table 7) provides information on the amount of time elapsed between the death event and sediment burial. To produce the degree of disarticulation, scavenging, and abrasion seen in the Agate bone bed would require a minimum interval of several months. The degree of abrasion on many bones indicates much more time is involved in some areas of the bone bed.

Age (mortality) profiles of the animals are informative as to the amount of time involved in the death event itself. Because sufficiently large samples of the small rhinoceros and the chalicothere are preserved in the Agate bone bed, the distribution of age classes can be worked out, using wear and eruption patterns of the teeth. If age classes of the fossils are distributed as in a modern standing ungulate population at a given point in time, then the death assemblage may have been derived from some instantaneous (or nearly so) death event, but if the sample displays other patterns of age class distribution, especially one suggestive of deaths resulting from time-averaged attritional sampling, then considerable time in the formation of the death assemblage may be involved.

The mortality profiles (based on complete lower dentitions) are the subject of a separate study (Hunt, in preparation) but the results can be briefly summarized. If we first assemble a mortality profile for a sample of 33 rhinoceroses (*Menoceras arikarense*) collected at random from sites scattered over the interchannel plains (volcaniclastic loess lithofacies), we find that it is made up primarily of very young and aged individuals (Fig. 27). This is the predicted age distribution that would be derived from a time-averaged sample of the population resulting from attritional deaths in the interchannel environment over a period of years. On the other hand, when a mortality profile is constructed for 140 rhinoceroses from the Agate bone bed at Carnegie Hill (Fig. 27), there is a strong representation of older prime adults (age classes 9–11), a good sampling of the younger subadults (age classes, 5–7) and very few aged animals. Within each age class, there is no clustering of crown height measurements around a class mode; thus, there is no indication that these rhinoceroses gave birth at approximately the same time, or died over a time interval sufficiently brief enough to preserve such clusters.

A word of caution: the sample from the Agate bone bed was largely derived from museum collections wherein the hundreds of Agate rhinoceroses are currently conserved. It is possible that the sample was biased by the style of field collecting in the early twentieth century, which often was not concerned with questions of population dynamics. In addition, once these fossils found their way into museums, many specimens were disposed of as unneeded duplicates, depleting the original sample, and possibly altering the relative proportion of different age classes.

However, the mortality profile for another Arikareean rhinoceros bone bed (77 Hill Quarry), situated on the northern edge of the Hartville Table 90 km (56 mi) northwest of the Agate Monument, closely approximates the profile obtained from Agate (Fig. 27). The depositional environment is extremely similar: the quarry samples a shallow waterhole setting filled by altered tuff nearly identical to that at Agate Monument. The 77 Hill profile was constructed from a population sample of a small dicerathere rhinoceros, collected by a field party of the American Museum and preserved intact in that institution. There is no reason to suspect significant bias in the collection of this sample, nor were specimens disposed of once the fossils were placed in the museum's collections. Again, as in the Agate profile, we observe a strong representation of older prime adults (age classes 9–11), a peak within the subadults (age classes 4–6), but with a few more aged individuals (age classes 12–14) than found at Agate. As at Agate, the profile does not suggest typical attritional mortality. The excessive representation of age classes 10–11 at Agate may be an artifact attributable to the durability of lower jaws of older animals in collections and during excavation.

TABLE 7. COMPARISON OF TAPHONOMIC FEATURES OF THE AGATE BONEBED (AT CARNEGIE HILL) AND HARPER QUARRY BONEBED, UPPER HARRISON BEDS, ARIKAREE GROUP, SIOUX COUNTY NEBRASKA

Agate Bonebed	Harper Quarry Bonebed
1. Large numbers of individuals of only two species of mammals.	1. Very few individuals of at least 12 species
2. Bonebed variable in thickness in different parts of waterhole.	2. Bonebed a monolayer of uniform thickness.
3. Bone density/m^2 varies from 0 to >150.	3. Bone density/m^2 varies from 0 to about 40.
4. Bones are generally disarticulated; residual articulations exist in some areas.	4. Bones entirely disarticulated; no residual articulations.
5. Behrensmeyer weathering stages 0-2 present; 94% of bones in stages 0-1; stage 2 very rare.	5. Behrensmeyer stages 0-2 present; most bones fall in stages 1-2; stage 2 more common than at Agate.
6. Bones near center of waterhole in unstable steeply plunging orientations associated with mammal tracks. Bones at periphery largely stable and flat lying.	6. 85% of bones plunge <20 degrees and lie in stable orientations; 92% plunge <30 degrees. No evidence of track bioturbation.
7. Bones of a single species may be segregated within the bonebed.	7. Bones of all species are randomly mixed within the bonebed.
8. High degree of skeletal representation in central part of waterhole; lesser representation at periphery.	8. Low degree of skeletal representation in bonebed.
9. Bone abrasion ranges from extreme to entirely unabraded. Abraded bone found near base of bonebed and at periphery of waterhole. Unabraded bones often common to skeletons in deeper part of waterhole.	9. Bone abrasion ranges from extreme to entirely unabraded. Much heavily abraded bone at base of bonebed. Unabraded bones mixed with abraded bones.
10. Selective abrasion patterns common to certain bones, especially astragalus, indicating oscillation on bed surface. Some bones heavily abraded.	10. Selective abrasion patterns not developed. Some bones heavily abraded.
11. Bones generally heavily scavenged, especially at waterhole periphery; central part of waterhole with some intact skeletons.	11. Bones generally heavily scavenged.
12. Trampling of sediments filling waterhole, resulting in some bone breakage and contorted bedding.	12. Trampling not in evidence but may have occurred.
13. Age profiles of rhinoceros suggest skeletons accumulated during a number of months but less than one year.	13. No species represented by sufficient numbers of individuals to construct age profiles.

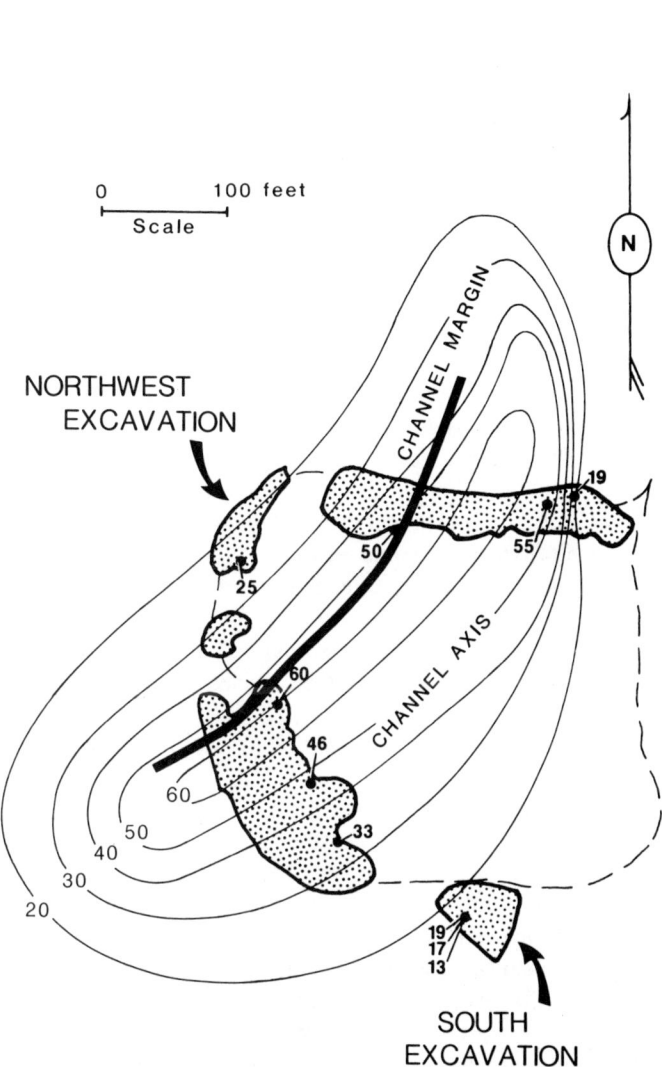

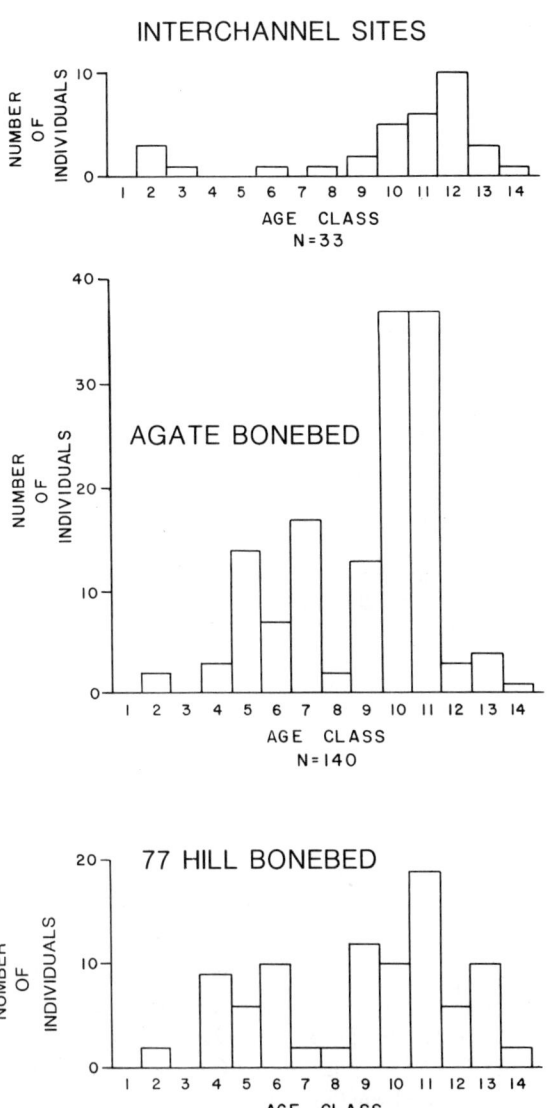

Figure 26. Plan map of the Carnegie Hill quarries (dot pattern) with superimposed contours showing the percentage of bones in test meters with a long axis plunging in excess of 10 degrees. Highest values are found in and near the deepest part of the channel axis along its west margin, where the bone mass is thickest, and track bioturbation is also most evident. Data for bones that plunge in excess of 20 degrees also approximate this same pattern. Closure on contours is not documented and is used to suggest only one of several possible contour patterns. Each percentage value is derived from excavation of 1 m^2 of bone bed; values west of the channel margin are from the upper bone bed; values east of the channel margin within the channel axis are from the lower bone bed.

Figure 27. Mortality profiles based on stage of dental eruption and wear in lower jaws of the small rhinoceroses *Menoceras* and *Diceratherium* from early Miocene Arikaree rocks of the Hartville Table. *Top,* random sampling of *Menoceras arikarense* from isolated sites in the interchannel volcaniclastic loess (lithofacies F2-MS), showing typical time-averaged attritional profile. *Middle,* possibly biased sampling of *Menoceras arikarense,* from the Agate bone bed (probably primarily from the lower bone bed but some jaws from the upper bone bed may be present), from a waterhole setting (lithofacies F1-CM). *Bottom,* random sampling of *Diceratherium* sp. from the 77 Hill bone bed, a waterhole setting with sediment infill nearly identical to the Agate bone bed, showing an unbiased mortality profile similar in general form to the profile obtained for *Menoceras* from the Agate bone bed, suggesting such profiles may be typical of waterhole deaths of Arikareean rhinoceroses during drought events in the North American midcontinent. Youngest age class is 1, oldest is 14. First lower molar erupted by age class 6; second lower molar erupted by age class 7; third molar by age class 9.

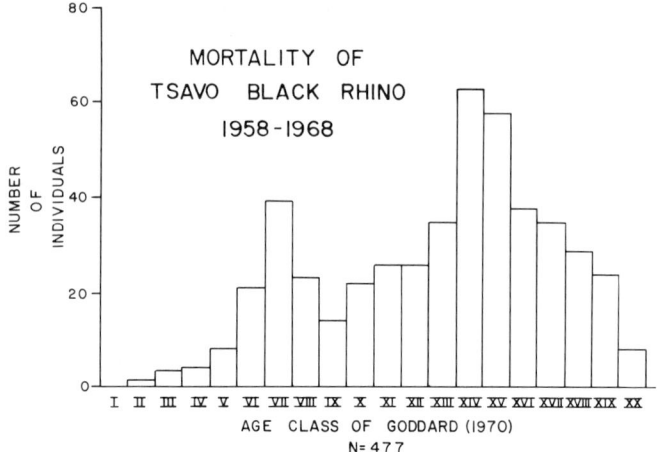

Figure 28. Mortality profile of living black rhinoceros *Diceros bicornis* in Tsavo National Park, Kenya, representing deaths from approximately 1958 to 1968, based on data from Goddard (1970). Note general similarity to mortality profiles of rhinoceroses from Agate Monument and 77 Hill bone beds (Fig. 27). Youngest age class is I, oldest is XX. First lower molar erupted by age class VI; second lower molar erupted by age class VIII, third molar by age class XI.

A most intriguing modern analogue to the Agate and 77 Hill mortality profiles is found in an attritional profile for the living black rhinoceros *Diceros bicornis* derived from data compiled by Goddard (1970). The profile (Fig. 28) is based on skulls collected during 1966 to 1968 within Tsavo National Park, Kenya; Goddard assumed that the collection represented natural mortality occurring between 1958 and 1968. This mortality profile was probably influenced by a severe drought in East Africa in 1961 (Foster, 1965): as a result of the drought, about 300 black rhinoceroses died in Tsavo Park west of the Yatta Plateau along the Athi River. Foster (1965) collected 49 of these skulls of drought-killed animals, and their age distribution is similar to that of the larger profile based on Goddard's data. The Goddard profile displays a bimodality crudely indicated in the Agate and 77 Hill profiles, and this bimodality in all three profiles includes a subadult peak, and an adult peak comprising age classes somewhat older than animals with freshly erupted third lower molar teeth. Subsequent work will be necessary to explore the similarities in these profiles more thoroughly. However, they tentatively suggest that the Agate and 77 Hill mortality profiles may result from cumulative deaths of primarily juvenile, subadult, and prime adult animals over a period of time in excess of several months, influenced by regional drought (87 to 96 percent of rhinoceroses in the 77 Hill and Agate profiles have reached only 60 percent or less of estimated lifespan, based on body size).

Underrepresentation of crania of very young rhinoceroses occurs in all three profiles (Figs. 27 and 28), and is attributed to the greater susceptibility of juvenile skulls to scavenging and weathering over time (Goddard, 1970, p. 113). At Agate, in our excavations, we have found bones of juveniles and even neonates in the bone bed, but intact skulls are very rare, both in the field and in collections. Given the fact that scavenging is well documented in the bone bed, the destruction of fragile crania of young animals by scavengers probably occurred at Agate.

Interpretive summary of Agate bone bed origin and sediment burial. There are several lines of evidence indicating that the four sedimentation units enclosing the lower and upper Agate bone beds formed over a prolonged period of time (a number of months to years), and that the lower and upper bone beds were the result of a death event also spanning an interval of time in excess of several months' duration.

(1) The lithofacies (F1–CM) that includes the lower and upper Agate bone beds contains mammal track horizons at two levels, penetrating downward from the surface of sedimentation units A and C (Fig. 19). Within the channel axis, these two track horizons are separated by an undisrupted bed of finely laminated fluvial sandstone. Units A and C, when wet, were sufficiently plastic to preserve track molds, yet when buried beneath fine-grained fluvial sand, they possessed a grain packing firm enough to allow only very limited penetration by the feet of mammals at higher levels, suggesting that they had dried and compacted. The presence of two track levels separated by an undisrupted compacted sandstone bed indicates that the waterhole fill was not rapidly deposited in one brief sedimentation event, but rather that pauses in sedimentation occurred, during which mammals bioturbated the bottom of the waterhole, and that the waterhole dried at least once as it filled.

(2) Two distinct bone layers were discovered during our 1986 field work in the thicker central part of the Carnegie Hill bone bed within the channel axis, confirming earlier unsubstantiated reports of a double bone layer at Carnegie Hill by Albert Thomson in 1918. The bone beds are found within the base of sedimentation units A and C. The upper bone bed may result from a second death event, but the presence of the same rhinoceros and chalicothere as in the lower bone bed, and the restriction of the upper bone bed to an area extending from the west bank of the channel a short distance eastward into the center of the waterhole, indicates that the upper bone level may result from redistribution of skeletal remains found on the west bank of the channel. These bones may have remained unburied on the channel bank after the first death event, and at a later time (following deposition of sedimentation unit B) were introduced into the waterhole by scavengers and other animals frequenting the site.

(3) Trampling of a rhinoceros skull within sedimentation unit C resulted in breakage and slight downward displacement of the upper jaw. However, neither the jaw nor the foot of the animal that broke the jaw penetrated downward 10 cm to disrupt finely laminated sandstone (unit B) intervening between the skull and sedimentation unit A. There is no doubt that sedimentation units A and B were compacted and resistant to penetration at that time, whereas unit C was viscous and mobile.

(4) In at least two locations, the surface of sedimentation unit A is incised by small fluvial scours a few inches in width and depth, producing sharply defined smooth-sided troughs, which filled with fluvial sand, demonstrating the compact cohesive na-

ture of unit A early in the filling of the waterhole. To cut such a trough requires that the sediment of unit A be firm, ranging from viscous to solid. In addition, the water level must have been very shallow to cut such small troughs. Within the sandy fill of adjacent track molds are also found angular pieces of white tuff of unit A floating in sandy matrix. Consequently, water level must have been very low, and in places the pond probably dried, indicating a pause in sedimentation.

(6) Age profiles of the small rhinoceros in the Agate bone bed suggest deaths distributed over a prolonged interval, on the order of months, as discussed earlier.

The deaths of rhinoceros and chalicothere at the Agate waterhole occurred prior to filling of the depression with fine-grained sediment. Before sedimentation was initiated, the carcasses underwent near-total dissociation, and were scavenged, particularly at the periphery of the waterhole. Disarticulation in warm, ponded water requires only a few weeks to months, and scavenging probably aided the disarticulation process. Abrasion of bones postdates scavenging of the carcasses, and can be pronounced at the base of the bone bed and at its periphery. The degree of abrasion indicates that a significant amount of time, on the order of months, intervened between death and final burial of these bones in the sediment of unit A. The presence of unusual patterns of selective abrasion on some bones demonstrates that they oscillated on the bed surface, and were thereby abraded without transport, held in position by the interlocking fabric of the bone bed, which lay like a carpet over the channel floor. The movement of water, sand, and bones in the waterhole produced by the feet of mammals coming to the site to drink probably played an important role in bone abrasion, but some abrasion is surely the result of streamflow through the bone bed prior to abandonment of the channel.

As the pond began to dry during a time of increasing aridity, volcanic ash and lime mud began to accumulate on the old channel floor, becoming mixed with sand at the west channel margin. The bone bed was gradually buried, particularly the deeper central areas of the waterhole. At the periphery, the falling water level exposed bones to subaerial weathering and additional scavenging and trampling. Other mammals, including huge entelodonts, trampled the bones buried within these fine sediments accumulating in the deeper parts of the waterhole.

A brief period of seasonal rainfall rejuvenated streams that introduced fine ash-rich sand into the western margin of the waterhole completely burying the lower bone bed. Over the remainder of the waterhole fine ash and lime mud continued to accumulate in the standing water that remained after the rains. Skeletal debris left on the west bank of the waterhole after the first death event was redistributed over the floor of the waterhole by scavengers to form a second or upper bone bed of lesser areal extent than the lower bone bed. The upper bone bed was gradually buried by fine sand and ash, and was trampled by mammals visiting the pond.

The following year, streams swollen with rainfall, marking initiation of the wet season, overtopped their banks and carried fine sand over the waterhole, burying both upper and lower bone beds. Following the rains, the waterhole dried and a soil developed over the bone deposit.

Comparison of the Agate and Harper Quarry bone beds. When the sum of the evidence from bone distribution, orientation, sorting, and abrasion—and from disarticulation, weathering, and animal activity—is considered, we can recognize the following similarities between the Agate and Harper Quarry bone beds: (1) both occur at exactly the same stratigraphic horizon (together with the American Museum–Cook Quarry) although separated by many miles, and include contemporaneous land mammals; (2) both include dense concentrations of mostly disarticulated bones within a well-defined local area; (3) the bones occur in fine-grained, ash-rich sediments filling shallow local depressions cut by streams in a stabilized land surface; (4) both include as a significant lithology a white calcareous tuff rich in ostracods, pulmonate gastropods, diatoms, charophyte, and other plant debris, indicating that when sediment filled these depressions, they were ponded environments without vigorous through-going streamflow; (5) both underwent a period of bone processing of some duration that intervened between the death of mammals at the sites and the initiation of sediment burial; this period of processing is recognized by nearly complete to complete skeletal disarticulation, preburial bone breakage attributable to scavenging and trampling, surface abrasion of bones, and to a lesser extent, weathering of bone surfaces (probably inhibited by immersion of bones in water at both localities).

The two bone beds differ in the following features: (1) at Agate the bone bed is dominated by two kinds of ungulates, whereas at Harper Quarry a wider diversity of species is present; (2) the degree of processing of the Harper Quarry bone bed is somewhat greater than at Agate in terms of abrasion, weathering, and amount of preburial breakage; (3) the Harper Quarry bone bed appears to display a greater degree of bone alignment due to flowing water than is evident at Agate; (4) bone density per square meter at Harper Quarry is most similar to the less dense peripheral aeras of the Agate bone bed, and does not attain the thickness or densities observed in the central part of the Agate waterhole; and (5) most bones in the Harper Quarry bone bed lie in stable orientations on the channel floor, in similar fashion to bones in the peripheral areas of the Agate bone bed; the unstable orientations found in the central bioturbated part of the Agate bone bed have not been found in Harper Quarry.

To discover these mammalian bone beds within terrestrial rock sequences is not unusual but is infrequent: to find that they occur at exactly the same stratigraphic level on a regional scale begs explanation. At three localities at which the F1–CM lithofacies developed, bone beds are present. The bones always occur in the base of the lithofacies, and have experienced a period of environmental processing before burial in fine-grained tuffaceous pond sediments. At all sites, bones are disarticulated, scavenged, and then abraded, resulting in skeletal residues differing in species diversity but sampling the same regional Late Arikareean fauna.

The bone-bearing lithofacies is deposited in abandoned

channel segments and depressions that incise the surface of the Harrison Formation. A regional unconformity and prominent silcrete paleosol (the oldest Arikaree silcrete known in the region) mark the top of the Harrison Formation, and record a long interval of nondeposition and soil formation throughout southeastern Wyoming and western Nebraska. The Upper Harrison channels containing the bone beds mark the beginning of a renewed regional episode of downcutting and subsequent aggradation. The considerable amount of volcanic ash in this lithofacies and throughout the Arikaree Group indicates the importance of volcanic activity to this aggradational process. In essence, the bone beds belong to the first sedimentation events that were preserved as part of a new cycle of regional aggradation. This final or "Upper Harrison" cycle is the culmination of Arikaree volcaniclastic sedimentation in the midcontinent, ending at about 19 Ma.

Severe seasonal drought accompanying a drier regional climate followed by renewed aggradation of fine volcaniclastics seems a compelling explanation for the origin of these bone beds. Drought is a mechanism likely to produce mass deaths of water-dependent mammals in a regional context, but one in which deaths might be prolonged over a number of months, yet still produce age profiles of the death assemblage similar to profiles expected as a result of sudden (catastrophic) events.

Living rhinoceroses are known to be water-dependent mammals (Ritchie, 1963), and the fossil record of chalicotheres demonstrates that they are most common in well-watered woodland environments. The teeth of chalicotheres suggest they were browsing mammals that may well have been limited to the stream courses. Their bones are common in upper Arikaree stream deposits but are extremely rare in interchannel environments. We know that the Agate Monument bone bed developed in a major paleovalley of the region, and that the Harper Quarry bone bed formed in a much smaller local drainage. We can visualize a scenario where mass death events of the larger water-dependent mammals were restricted to their customary habitat, the principal valleys of the tableland. As drought intensified, rhinoceros and chalicothere were drawn to the last major water sources in these valleys. Gradually, over a period of a few months, as all available vegetation was consumed in the vicinity of these water sources, these mammals expired in small groups in and near the waterholes. The taphonomic history of the Agate bone bed is in accord with a waterhole whose margins were heavily worked by scavengers, but in which other areas were largely undisturbed, either because scavengers moved on to more hospitable environments or, satiated, they ignored the last groups to die within the waterhole environment. Mass deaths in which scavengers have not touched large numbers of carcasses are known from recent plains environments in East Africa (Schaller, 1972).

On the other hand, Harper Quarry exploits a bone bed that formed along a small local ephemeral stream some distance from the principal valleys of the region. Drought brought a varied group of mammals to the site, but no herds of water-dependent mammals were trapped there. The skeletal debris in the bone bed represents normal attritional deaths at the waterhole, probably supplemented by additional individuals drawn to the site that fell victim to the drought conditions.

CARNIVORE DEN COMMUNITY

In 1981 a large den complex (Fig. 29) containing the skeletal remains of at least five species of mammalian carnivores was discovered about 150 m (500 ft) from the main Agate bone bed at Carnegie Hill (Hunt and others, 1983). These dens are burrowed into the sediment fill of the waterhole; thus, they postdate the Carnegie Hill death assemblage by some unknown but geologically brief amount of time. The dens have produced remains of two species of amphicyonid beardogs, two species of mustelids, and a small canid. The site is the oldest carnivoran den complex known in the fossil record, and allows a rare glimpse into the ecology of the carnivore community during this time interval. The dens were situated within the stream valley, near water and potential prey. The only mammal bones other than Carnivora found in association with these dens are rare bones of juvenile ungulates such as rhinoceros, camel, and oreodont, suggesting that young animals were a major food resource.

The beardog *Daphoenodon,* the largest carnivore found in the dens, is the size of a wolf. The only larger carnivore known from this time interval is the beardog *Ysengrinia,* about the size of a large black bear (Hunt, 1972). Although not found in the dens, it occurs in the Agate bone bed and other nearby sites. Whereas some ungulate mammals (chalicothere, entelodont, dicerathere rhinoceros) in the community were considerably larger than *Ysengrinia,* most others (camels, horses, oreodonts, small rhinoceros, deerlike ruminants, peccary) were about the same size or smaller. No mammal in the community exceeded 1,000 kg, whereas in modern and Pleistocene communities, mammals more than 1,000 kg are not uncommon (rhinoceros, proboscideans, hippopotamus, giraffids). In keeping with the smaller body size of the largest ungulates, we do not find any truly large carnivorans (>200 kg) in the Upper Harrison mammal fauna.

Among the mid-sized members (10 to 80 kg) of the carnivore community are a few species that show early postcranial specializations for an open-country habitus. Two species of beardogs, an ursid, and a canid were at least moderately cursorial, as evidenced by limb structure, but were not as specialized for this role as living cursorial canids. There is a suggestion that the beardog *Daphoendon* could have been gregarious, and hunted in packs, much like wolves.

However, several of the larger Upper Harrison carnivores (*Ysengrinia, Megalictis*) are short-footed and short-legged, with lower limb segments equal to or shorter than the upper limb segments. This limb anatomy is accompanied by powerful musculature. Such carnivores seem best adapted for a method of prey capture that utilizes ambush, and that requires strength to win the struggle, and effect the kill. These animals have been found only in the sediments of the stream valleys and not in the interchannel

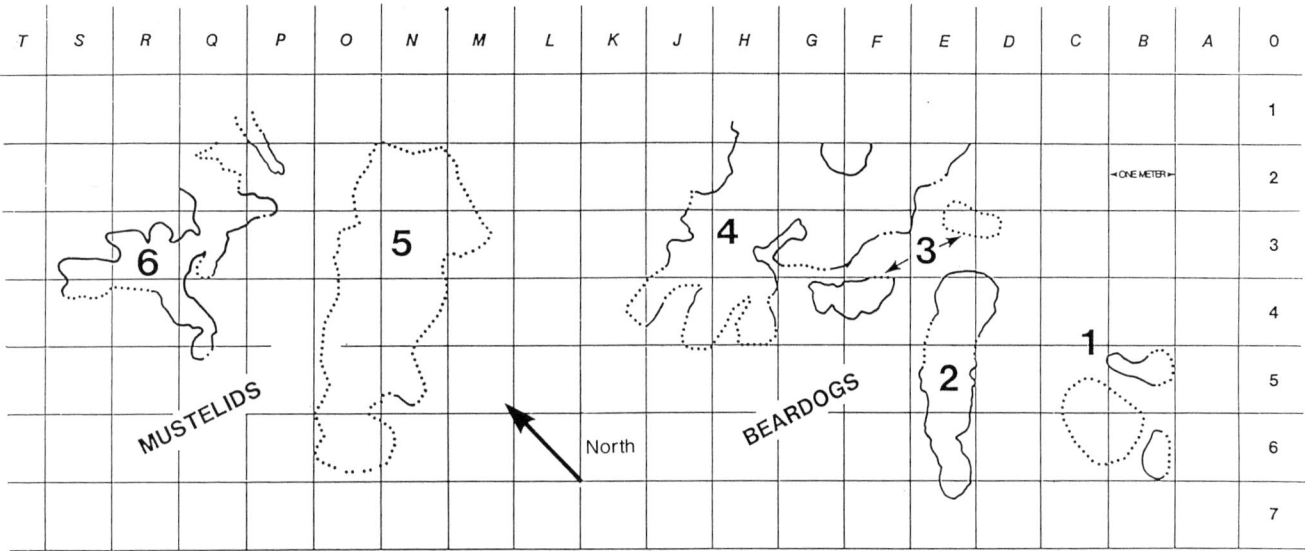

Figure 29. Plan map of carnivore dens discovered by University of Nebraska field parties between 1981 and 1985 at Carnegie Quarry 3, Beardog Hill, Agate Fossil Beds National Monument. The dens are excavated in tuffaceous bedrock and were subsequently filled by fluvial ash-rich sands that buried many animals within the burrows. Dens 1 through 4 and vicinity have produced mostly skeletal remains of extinct beardogs (Amphicyonidae). Den areas 5 and 6 yielded many bones of mustelid carnivores but very few of beardogs. Dotted lines indicate uncertain boundaries of burrows; solid lines are established burrow perimeters. The Quarry 3 dens represent the oldest evidence of denning behavior by large carnivorous mammals known in the fossil record.

setting, and may have lived primarily in the more densely vegetated stream-border environment.

The smaller carnivores (<10 kg) are canids (*Tomarctus, Phlaocyon*) and mustelids (*Promartes*) that were likely generalists, taking both small vertebrate prey, invertebrates, and plant material in season. Thus the carnivore assemblage as a whole seems compatible with well-vegetated environments providing cover, as well as more open settings in which semicursorial species could operate effectively, perhaps even in social hunting groups.

Both carnivores and their ungulate prey, when evaluated as a fauna, include species that would seem best adapted to densely vegetated terrain, as well as others that appear to be on the verge of developing effective anatomical specializations for more open country. This agrees with other evidence in indicating well-vegetated stream-border communities, spatially separated by open interchannel reaches occupied by grassland, or possibly low shrub savanna or open savanna–parkland environments (for definition of savanna environments, see Cole, 1986).

RELATION BETWEEN VOLCANICLASTIC SEDIMENTATION AND THE PRESERVATION OF UPPER ARIKAREE TERRESTRIAL PALEOCOMMUNITIES

Without the influx of enormous quantities of volcanic ash into the midcontinent during the early Miocene, the mammalian paleocommunities of stream valley and interchannel plains would not have been preserved in the North American fossil record (Fig. 30). Mid-latitude westerly winds transported Arikaree sediment eastward into the continental interior, mixing pyroclastics with a small but persistent epiclastic fraction from Rocky Mountain uplifts. The allochthonous volcaniclastics accumulated in a semiarid to arid frost-free climate in which alternating wet and dry seasons encouraged the development of open grasslands, and limited the development of woodland to the stream-border environment. Grassland vegetation probably played an important role in the trapping of pyroclastic debris, and in the development of level regional geomorphic surfaces.

Aridity within the continental interior, together with subsurface waters of near-neutral pH, restricted diagenetic alteration of ash-rich fine-grained Arikaree volcaniclastics to slight alteration and dissolution of shards and unstable mineral grains, and the development of pore-filling cements (Stanley and Benson, 1979). The high ash content of these Arikaree sandstones and the absence of significant diagenetic alteration of the rock fabric significantly contributed to their marked porosity and permeability.

Pronounced diagenesis of Arikaree sediments is restricted primarily to paleosol horizons. Paleosols useful in the demonstration of regional paleotopography are traceable for many miles only because of the widespread geographic distribution of the volcaniclastic loess lithofacies. It is the paleosols that provide the confirming evidence for the existence of flat regional topographic surfaces, the interchannel plains, east of the Rocky Mountain front at this time in the early Miocene.

Shallow, wide, sandy, ephemeral to intermittent streams of

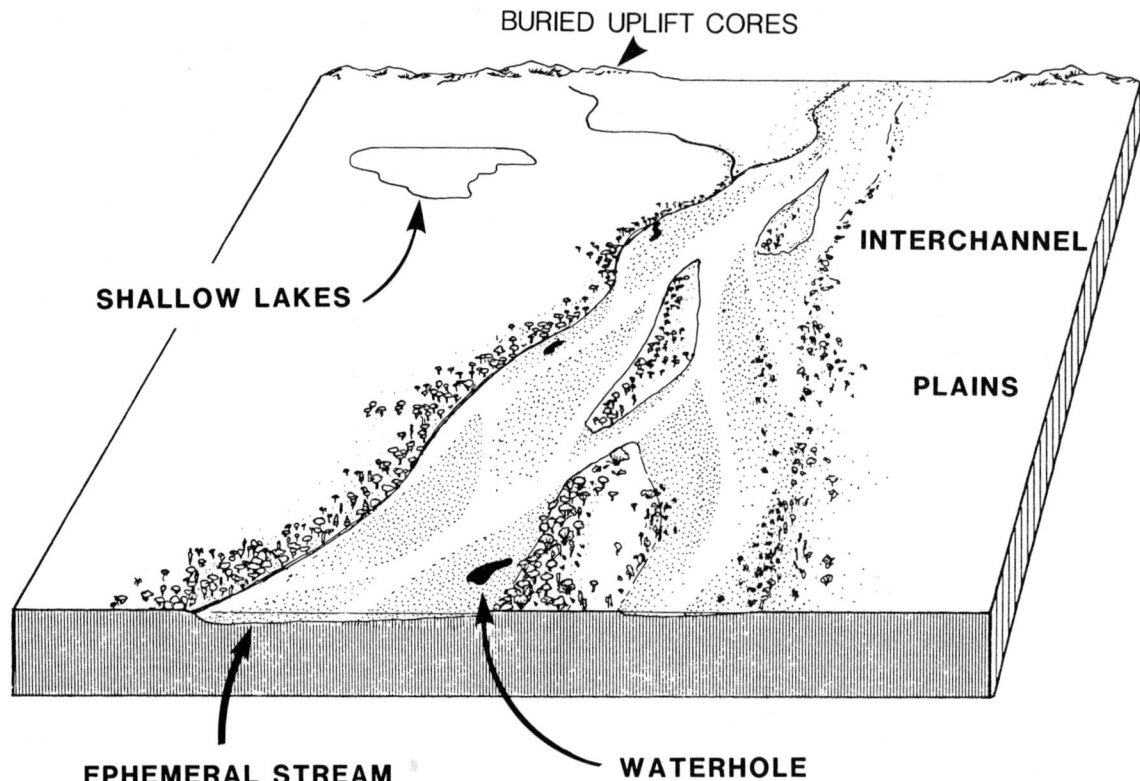

Figure 30. Diagram of early Miocene paleoenvironments east of the Hartville Uplift of the Rocky Mountains, based on combined data from upper Arikaree Group lithofacies and biofacies. Volcanic activity in western North America produced large volumes of pyroclastic ash that mantled level plains and stream valleys within the semiarid continental interior. Paleotopographic relief between streams and adjacent plains was subdued. Stream-border savanna-woodland or savanna-parkland associated with wide shallow valleys provided vegetative cover and food resources for browsing and grazing mammals. Water was probably seasonally available, found during times of rainfall in channels and pools along intermittent or ephemeral stream courses, and in shallow lakes on the open plains. The interchannel plains, which were drier than the stream-border environment, supported grassland, possibly with a few sparse shrubs or low trees. Warm, shallow, holomictic lakes were present in the interchannel reaches, and persisted as the only evident source of water in the region when the drainages became filled with volcaniclastic sediments and streamflow had ceased. Periodic drought forced water-dependent mammalian species to ephemeral waterholes where lack of forage led to mass deaths extending over intervals of months, whereas a number of drought-resistant species seem to have been little affected by dry conditions.

low gradient and sinuosity traversed the plains, early in their history reworking a fine-grained volcaniclastic bedload, and later eventually filling with fine loessic volcaniclastic sediment. Epiclastic sediment makes only a very minor contribution to these valley fills. Coarse crystalline debris from nearby uplifts is conspicuously absent from these streams, and indicates the burial of adjacent ranges by a blanket of fine-grained volcaniclastic loess.

Lacustrine limestones indicative of ephemeral, shallow, holomictic lakes occupied depressions on the plains, in some cases extending over several square miles. The lakes contained a depauperate biota of ostracods, aquatic pulmonate gastropods, and perilacustrine vegetation, and are lacking in aquatic vertebrates, especially fish. Almost all of these lacustrine limestones are enclosed within volcaniclastic loess, and would not have been preserved without the influx of volcanic detritus into the region.

Mammalian fossils preserved in the interchannel volcaniclastic loess and in the stream channel and flood-plain units are largely attritional samples of ungulates and occasional carnivores. Bones are commonly scattered, weathered, and scavenged; they were subjected to limited fluvial transport and dispersion by predators/scavengers in the stream valleys, but were transported (dispersed) only by scavenging mammals in the interchannel plains environment.

Bone beds occur in waterhole settings in the stream valleys where regional drought seems to be the agency responsible for the large numbers of individuals gathered together at these death sites. Disruption in the cycling of wet and dry seasons stressed water-dependent mammals at times of regional drought. Taphonomic reconstructions of the Agate and Harper Quarry bone beds serve as models for such sites. Age (mortality) profiles derived from mass death assemblages of rhinoceros and chalicothere in these bone beds suggest the death assemblages developed over a period of months, and subsequent filling of the waterholes with fine-grained tuffaceous sediments took an equivalent amount of time. Without the volume of volcaniclastic sediment introduced into these environments, the mammalian fossil record of the various paleocommunities would be much more poorly indicated, and certain environments such as the vast interchannel reaches would not be represented at all.

The mammalian carnivore community was made up entirely of amphicyonids (beardogs), mustelids, canids, and an ursid. There are no cats of any kind, and this void was filled by development of marked diversity among the arctoid carnivorans. Skeletal remains of amphicyonids, mustelids, and canids have been found in burrows making up a large den complex excavated in fine-grained volcaniclastic sandstone. The den community was situated within a shallow, wide, aggrading stream valley at Agate Fossil Beds National Monument, and in proximity to the waterhole environment that preserved the Agate bone bed. The use of burrows as shelters by many of these carnivores indicates that denning was important in the ecology of the predator community. Preservation of a broad age spectrum of carnivores in the dens, and of multiple species, suggests a sudden death event or series of such events, probably flash flooding that drowned a number of animals in the burrows. The fact that the burrows are filled by, and are also excavated in, fine tuffaceous sediment again emphasizes the importance of volcaniclastics in the preservation of the biota.

It is the prolonged yet episodic introduction of fine-grained volcaniclastic detritus into a semiarid to arid continental interior that provided the essential conditions for preservation of land mammals in all major environments. Although the seasonally dry climate did not prevent decay and disintegration of bone, the steady input of sediment ensured burial of some of this skeletal material. Widespread distribution of loessic volcaniclastics in these environments preserved a sample of the death assemblages in each of these settings, allowing inferences to be drawn about the biocoenoses that actually existed in them in the past. The combination of porous, tuffaceous, fine-grained sediments and pH-neutral to mildly alkaline subsurface waters contributed to the preservation of vertebrate and invertebrate skeletal material. Such faunal representation of both interchannel and streamborder environments on a regional scale is not usually encountered in the terrestrial Cenozoic record, much of which was deposited in geographically restricted fluvial channel and floodplain settings.

ACKNOWLEDGMENTS

I wish to express my appreciation to Martin Lockley for his invitation to contribute a paper to the Symposium on Volcanism and Fossil Biotas. I also thank Marc Marcuson for many of the illustrations, J. A. Breyer for the textural analyses of Figure 9, and J. A. Breyer and R. I. Skolnick for petrographic counts of thin sections and grain mounts in Tables 2 and 4. I benefited from x-ray diffraction analyses of intergranular cements kindly carried out by M. A. Holmes. Various aspects of the study were supported by the National Geographic Society and National Science Foundation. I am grateful to the National Park Service for the opportunity to excavate at Agate Fossil Beds National Monument, and particularly appreciate the cooperation of Superintendents A. J. Banta and J. Kyral, and Unit Manager John Rapier, who facilitated our work in numerous ways.

REFERENCES CITED

Behrensmeyer, A. K., 1978, Taphonomic and ecologic information from bone weathering: Paleobiology, v. 4, p. 150–162.

Binford, L. R., 1981, Bones; Ancient men and modern myths: New York, Academic Press, 320 p.

Bishop, W. W., 1963, The later Tertiary and Pleistocene in eastern equatorial Africa, in Howell, F. C., and Bourliere, F., eds., African Ecology and Human Evolution: London, Methuen and Co., p. 246–275.

Bishop, W. W., 1968, The evolution of fossil environments in East Africa: Leicester Literary and Philosophical Society Transactions, v. 62, p. 22–44.

Brain, G. K., 1981, The hunters or the hunted?: An introduction to African cave taphonomy: Chicago, Illinois, University of Chicago Press, 365 p.

Burbank, D. W., and Li Jijun, L., 1985, Age and paleoclimatic significance of the loess of Lanzhou, North China: Nature, v. 316, p. 429–431.

Calkins, F. C., 1902, A contribution to the petrography of the John Day Basin, Oregon: University of California Publications, Department of Geology Bulletin 3, p. 109–172.

Cole, M., 1986, The savannas; Biogeography and geobotany: New York, Academic Press, 438 p.

Coombs, M. C., 1975, Sexual dimorphism in chalicotheres (Mammalia, Perissodactyla): Systematic Zoology, v. 24, p. 55–62.

Darton, N. H., 1899, Preliminary report on the geology and water resources of Nebraska west of the 103rd meridian: U.S. Geological Survey 19th Annual Report 1897–98; Part 4, Hydrography: U.S. Geological Survey, p. 719–785.

Diffendal, R. F., 1982, Regional implications of the geology of the Ogallala Group (Miocene), Banner, Kimball, and Morrill Counties, Nebraska, and adjacent areas: Geological Society of America Bulletin, v. 93, p. 964–976.

—— , 1984, Comments on the geologic history of the Ogallala Formation in the southern panhandle of Nebraska, in Whetstone, G., ed., Proceedings of the Ogallala Aquifer Symposium 2: p. 194–216.

Emry, R., Bjork, P., and Russell, L., 1987, The Chadronian, Orellan, and Whitneyan North American land mammal ages, in Woodburne, M. O., ed., Cenozoic mammals of North America: Berkeley, University of California Press, p. 118–152.

Fisher, R. V., and Rensberger, J. M., 1972, Physical stratigraphy of the John Day Formation, central Oregon: University of California Publications in the Geological Sciences, v. 101, p. 1–45.

Fisher, R. V., and Schmincke, H. U., 1984, Pyroclastic rocks: Berlin, Springer-Verlag, 472 p.

Foster, J. B., 1965, Mortality and aging of black rhinoceros in East Tsavo Park, Kenya: East African Wildlife Journal, v. 3, p. 118–119.

Freytat, P., and Plaziat, J.-C., 1982, Continental carbonate sedimentation and pedogenesis; Late Cretaceous and early Tertiary of southern France: Contributions to Sedimentology, v. 12, p. 1–213.

Gilbert, G. K., 1896, Underground water of the Arkansas Valley in eastern Colorado: U.S. Geological Survey Annual Report, v. 17, p. 25.

Goddard, J., 1970, Age criteria and vital statistics of a black rhinoceros population: East African Wildlife Journal, v. 8, p. 105–121.

Goodwin, R., and Diffendal, R. F., 1987, Paleohydrology of some Ogallala (Neogene) streams in the southern panhandle of Nebraska, in Ethridge, F., Flores, R., and Harvey, M., eds., Recent developments in fluvial sedimentology; Contributions from the 3rd International Fluvial Sedimentology Conference: Society of Economic Paleontologists and Mineralogists Special Publication 39, p. 149–157.

Gustavson, T. C., and Winkler, D. A., 1988, Depositional facies of the Miocene–Pliocene Ogallala Formation, northwestern Texas and eastern New Mexico: Geology, v. 16, p. 203–206.

Halffter, G., and Matthews, E., 1966, The natural history of dung beetles of the subfamily Scarabaeinae: Folia Entomologica Mexicana, v. 12–14, p. 1–312.

Hallam, A., 1981, Facies interpretation and the stratigraphic record: San Francisco, California, W. H. Freeman and Co., 291 p.

Hatcher, J. B., 1893, The Titanotherium beds: American Naturalist, v. 27, p. 204–221.

—— , 1902, Origin of the Oligocene and Miocene deposits of the Great Plains: Proceedings of the American Philosophical Society, v. 41, p. 113–131.

—— , 1903, Narrative of the Princeton University expeditions to Patagonia, March 1896 to September 1899: Princeton, New Jersey, Reports of the Princeton University Expeditions to Patagonia 1 (1896–1899), 314 p.

Haworth, E., 1897, Physical properties of the Tertiary: Geological Survey of Kansas Report 2, p. 247–284.

Hay, R. L., 1963, Stratigraphy and zeolitic diagenesis of the John Day Formation of Oregon: University of California Publications in the Geological Sciences, v. 42, no. 5, p. 199–262.

—— , 1976, Geology of the Olduvai Gorge: Berkeley, University of California Press, 203 p.

—— , 1986, Role of tephra in the preservation of fossils in Cenozoic deposits of East Africa, in Frostick, L. E. and others, eds., Sedimentation in the African rifts: Geological Society of London Special Publication 25, p. 339–344.

Hayden, F. W., 1857, Notes on the geology of the Mauvaises Terres of White River, Nebraska: Philadelphia, Pennsylvania, Proceedings of the Academy of Natural Sciences, v. 9, p. 151–158.

Haynes, G., 1980, Prey bones and predators; Potential ecologic information from analysis of bone sites: Ossa, v. 7, p. 75–97.

—— , 1982, Utilization and skeletal disturbances of North American prey carcasses: Arctic, v. 35, no. 2, p. 266–281.

Hill, A., 1979, Disarticulation and scattering of mammal skeletons: Paleobiology, v. 5, no. 3, p. 261–274.

Hill, A., and Behrensmeyer, A. K., 1984, Disarticulation patterns of some modern East African mammals: Paleobiology, v. 10, no. 3, p. 366–376.

Hunt, R. M., 1972, Miocene amphicyonids (Mammalia, Carnivora) from the Agate Spring Quarries, Sioux County, Nebraska: American Museum Novitates 2506, p. 1–39.

—— , 1978, Depositional setting of a Miocene mammal assemblage, Sioux County, Nebraska: Palaeogeography, Palaeoclimatology, Palaeoecology, v. 24, p. 1–52.

—— , 1981, Geology and vertebrate paleontology of the Agate Fossil Beds National Monument and surrounding region, Sioux County, Nebraska (1972–1978): National Geographic Society Research Reports 13, p. 263–285.

—— , 1985, Faunal succession, lithofacies, and depositional environments in Arikaree rocks (lower Miocene) of the Hartville Table, Nebraska and Wyoming, in Martin, J. E., ed., Fossiliferous Cenozoic deposits of western South Dakota and northwestern Nebraska: Dakoterra, v. 2, no. 2, p. 155–204.

Hunt, R. M., Xue, X. X., and Kaufman, J., 1983, Miocene burrows of extinct beardogs; Indication of early denning behavior of large mammalian carnivores: Science, v. 221, p. 364–366.

Johannsen, A., 1914, Petrographic analysis of the Bridger, Washakie, and other Eocene formations of the Rocky Mountains: Bulletin of the America Museum of Natural History, v. 33, p. 209–222.

Lajoie, J., 1984, Volcaniclastic rocks, in Walker, R. G., ed., Facies models, 2nd ed.: Geoscience Canada Reprint Series 1, p. 39–52.

Liu, T., 1985, Loess in China, 2nd ed.: Berling, Springer-Verlag, 224 p.

Marshall, L. G., 1985, Geochronology and land mammal biochronology of the Transamerican faunal interchange, in Stehli, F., and Webb, S. D., eds., The great American biotic interchange: Topics in Geobiology, v. 4, p. 49–58.

Marshall, L. G., Hoffstetter, R., and Pascual, R., 1983, Mammals and stratigraphy; Geochronology of the continental mammal-bearing Tertiary of South America: Montpellier, France, Palaeovertebrata Mémoire Extraordinaire 1983, p. 1–93.

Matthew, W. D., 1899a, A provisional classification of the freshwater Tertiary of the West: Bulletin of the American Museum of Natural History, v. 12, p. 19–75.

—— , 1899b, Is the White River Tertiary an eolian deposit?: American Naturalist, v. 33, p. 403–408.

—— , 1901, Fossil mammals of the Tertiary of northeastern Colorado: American Museum of Natural History Memoir 1, p. 353–447.

—— , 1915, The Tertiary sedimentary record and its problems, in Schuchert, G., ed., Problems of American geology: New Haven, Connecticut, Yale University Press, p. 377–478.

—— , 1923, Fossil bones in the rock: Natural History, v. 23, no. 4, p. 358–369.

Meek, F. B., and Hayden, F. V., 1861, Descriptions of new Lower Silurian, Jurassic, Cretaceous, and Tertiary fossils, collected in Nebraska, by the exploring expedition under the command of Capt. Wm. F. Reynolds: Philadelphia, Pennsylvania, Proceedings of the Academy of Natural Science, v. 13, p. 415–447.

Merriam, J. C., 1901, A contribution to the geology of the John Day Basin, Oregon, California University Publications, Department of Geology Bulletin, v. 2, p. 269–314.

Naeser, C. W., Izett, G. A., and Obradovich, J. D., 1980, Fission-track and K-Ar ages of natural glasses: U.S. Geological Survey Bulletin 1489, p. 1–31.

Obradovich, J. D., Izett, G. A., and Naeser, C. W., 1973, Radiometric ages of volcanic ash and pumice beds in the Gering Sandstone (earliest Miocene) of the Arikaree Group, southwestern Nebraska: Geological Society of America Abstracts with Programs, v. 5, p. 499–500.

Pascual, R., Vucetich, M., Scillato-Yane, G., and Bond, M., 1985, Main pathways of mammalian diversification in South America, in Stehli, F., and Webb, S. D., eds., The great American biotic interchange: Topics in Geobiology, v. 4, p. 219–247.

Patterson, B., and Pascual, R., 1972, The fossil mammal fauna of South America, in Keast, A., Erk, F., and Glass, B., eds., Evolution, mammals, and southern continents: Albany, State University of New York Press, p. 247–309.

Peterson, O. A., 1905, Description of new rodents and discussion of the origin of Daemonelix: Carnegie Museum Memoir 2, p. 139–191.

—— , 1907, The Miocene beds of western Nebraska and eastern Wyoming and their vertebrate faunae: Carnegie Museum Annals, v. 4, no. 1, p. 21–72.

—— , 1909, A revision of the Entelodontidae: Carnegie Museum Memoir 4, no. 3, p. 41–158.

—— , 1923, A fossil-bearing slab of sandstone from the Agate Spring Quarries of western Nebraska exhibited in the Carnegie Museum: Carnegie Museum Annals, v. 15, no. 1, p. 91–93.

Picard, M. D., and High, L. R., 1973, Sedimentary structures of ephemeral streams: Developments in Sedimentology, v. 17, p. 1–223.

Pickford, M., 1986, Sedimentation and fossil preservation in the Nyanza rift system, Kenya, in Frostick, L. E., and others, eds., Sedimentation in the African rifts: Geological Society of London Special Publication 25, p. 345–362.

Pye, K., 1987, Aeolian dust and dust deposits: New York, Academic Press, 334 p.

Reineck, H. E., and Singh, I. B., 1980, Depositional sedimentary environments:

New York, Springer-Verlag, 549 p.

Retallack, G. J., 1983, Late Eocene and Oligocene paleosols from Badlands National Park, South Dakota: Geological Society of America Special Paper 193, 82 p.

Ritchie, A. T., 1963, The black rhinoceros (*Diceros bicornis*): East African Wildlife Journal, v. 1, p. 54–62.

Rogers, A. F., 1924, Mineralogy and petrography of fossil bone: Geological Society of America Bulletin, v. 35, p. 535–556.

Russo, A., Flores, M., and Benedetto, H. D., 1980, Patagonia Austral Extraandina, *in* Segundo Simposio de Geología Regional Argentina, Córdoba, September 1976, v. 2: Córdoba, Argentina, Academia de Ciencias Naturales, p. 1431–1462.

Schaller, G. B., 1972, The Serengeti lion; A study of predator–prey relations: Chicago, Illinois, University of Chicago Press, 480 p.

Schmid, R., 1981, Descriptive nomenclature and classification of pyroclastic deposits and fragments: Geology, v. 9, p. 41–43.

Simpson, G. G., 1984, Discoverers of the Lost World: New Haven, Connecticut, Yale University Press, 222 p.

Sinclair, W. J., 1906, Volcanic ash in the Bridger beds of Wyoming: Bulletin of the American Museum of Natural History, v. 22, p. 273–280.

—— , 1909, The Washakie, a volcanic ash formation: Bulletin of the American Museum of Natural History, v. 26, p. 25–27.

—— , 1912, Contributions to geologic theory and method by American workers in vertebrate paleontology: Geological Society of America Bulletin, v. 23, p. 262–266.

Sinclair, W. J., and Granger, W., 1911, Eocene and Oligocene of the Wind River and Bighorn Basins: Bulletin of the American Museum of Natural History, v. 30, p. 83–117.

Skolnick, R. I., 1985, Geology and paleontology of the Lay Ranch beds, Sioux County, Nebraska, and Goshen County, Wyoming [M.S. thesis]: Lincoln, University of Nebraska, 156 p.

Smith, G. A., 1986, Coarse-grained nonmarine volcaniclastic sediment; Terminology and depositional process: Geological Society of America Bulletin, v. 97, p. 1–10.

Stanley, K. O., and Benson, L. V., 1979, Early diagenesis of High Plains Tertiary vitric and arkosic sandstone, Wyoming and Nebraska: Society of Economic Paleontologists and Mineralogists Special Publication 26, p. 401–423.

Suthren, R. J., 1985, Facies analysis of volcaniclastic sediments; A review, *in* Benchley, P., and Williams, B., eds., Sedimentology; Recent developments and applied aspects: London, Blackwell Scientific Publications, p. 123–146.

Swinehart, J., Souders, V., Degraw, H., and Diffendal, R., 1985, Cenozoic paleogeography of western Nebraska, *in* Flores, R., and Kaplan, S., eds., Paleogeography of the west-central United States: Rocky Mountain Section, Society of Economic Paleontologists and Mineralogists, p. 209–229.

Tedford, R. H., Swinehart, J., Hunt, R., and Voorhies, M., 1985, Uppermost White River and lowermost Arikaree rocks and faunas, White River valley, northwestern Nebraska, and their correlation with South Dakota, *in* Martin, J. E., ed., Fossiliferous Cenozoic deposits of western South Dakota and northwestern Nebraska: Dakoterra, v. 2, no. 2, p. 335–352.

Tedford, R. H., and others, 1987, Faunal succession and biochronology of the Arikareean through Hemphillian interval (late Oligocene through earliest Pliocene epochs) in North America, *in* Woodburne, M. O., ed., Cenozoic mammals of North America: Berkeley, University of California Press, p. 153–210.

Trimble, D. E., 1980, Cenozoic tectonic history of the Great Plains contrasted with that of the southern Rocky Mountain: A synthesis: Mountain Geologist, v. 17, no. 3, p. 59–69.

Vicars, R. G., and Breyer, J. A., 1981, Sedimentary facies in air-fall pyroclastic debris, Arikaree Group (Miocene), northwest Nebraska, U.S.A.: Journal of Sedimentary Petrology, v. 51, no. 3, p. 909–921.

Wanless, H. R., 1922, Lithology of the White River sediments: Philadelphia, Pennsylvania, Proceedings of the American Philosophical Society, v. 61, p. 184–203.

—— , 1923, The stratigraphy of the White River beds of South Dakota: Philadelphia, Pennsylvania, Proceedings of the American Philosophical Society, v. 62, p. 190–269.

Wortman, J. L.,1983, On the divisions of the White River or lower Miocene of Dakota: Bulletin of the American Museum of Natural History, v. 5, p. 95–105.

MANUSCRIPT ACCEPTED BY THE SOCIETY JUNE 21, 1989

Printed in U.S.A.

Evidence of flora and fauna in the gardens and cultivated land destroyed by Vesuvius in A.D. 79

Wilhelmina F. Jashemski
Department of History, University of Maryland, College Park, Maryland 20742

ABSTRACT

The destruction of the cities of Pompeii and Herculaneum and the villas in the surrounding area by the eruption of Vesuvius in A.D. 79 preserved a unique body of information about flora and fauna, such as is available at no other ancient site. Root cavities can be emptied of lapilli and casts made of roots that were growing at the time of the eruption. Planting patterns also help to identify the plants raised. Soil contours, which vary according to the crop, are preserved by the lapilli just as they were in A.D. 79. Carbonized, or partially carbonized fruit, nuts, vegetables, root material, branches, and twigs are at times preserved. Ancient pollen also furnishes valuable evidence. The bones of mammals and the shells of mollusks and echinoderms are often found, as well as a few bird bones. Paintings preserve the actual appearance of the ancient plants, mammals, birds, and marine animals. Unfortunately, much of this evidence has been lost, for until recently excavators have not been interested in such finds. Today scientific garden archaeology carefully salvages all such evidence. The excavated evidence is then studied, together with the comments of the ancient writers and in the light of modern practices, which strikingly continue those of antiquity. Many scientists, in diverse disciplines, have cooperated in identifying and interpreting this valuable evidence.

This chapter briefly reports the excavation of six different types of sites, three at Pompeii and three at nearby villas, and discusses specific examples of the various types of evidence of flora and fauna preserved by Vesuvius at these sites.

INTRODUCTION

The eruption of Vesuvius in August A.D. 79, which suddenly destroyed the flourishing cities of Pompeii and Herculaneum, as well as countless villas in the surrounding countryside, made the area a unique source of information about many aspects of ancient Roman life. Of great importance is the evidence for ancient gardens, and the flora and fauna found in them. At many sites archaeologists have uncovered public buildings, and occasionally houses or blocks of houses. But only at Pompeii do we have a complete city preserved and, in good part, excavated (Fig. 1). Only at Pompeii is it possible to walk up and down miles of streets and visit the homes, gardens, and shops of thousands of inhabitants. At Herculaneum, which is much more difficult to excavate because of the nature of the volcanic deposit that covered it, only four *insulae* (city blocks) and parts of four or five others have been excavated.

At the time of the eruption, a 17-year-old boy looked across the Bay of Naples and saw a strange cloud, shaped like an umbrella pine, rising from Vesuvius. Young Pliny was visiting his uncle, the famed natural scientist Pliny the Elder, who was commander of the Roman fleet stationed at Misenum on the Bay of Naples. Years later Pliny the Younger, in response to inquiries from the historian Tacitus, wrote two letters (Pliny, *Epistulae,* book 6, letters 16, 20) giving his remarkable eye-witness description of the catastrophic eruption of Vesuvius in which his uncle lost his life while trying to rescue friends. The term "Plinian eruption" used by volcanologists to describe all eruptions similar to that of Vesuvius in A.D. 79 is derived from the name of Pliny the Younger. The important recent study of Sigurdsson and others (1985) describes the stratigraphy of the volcanic deposits at the various Vesuvian sites and relates them to the eye-witness

---

Jashemski, W., 1990, Evidence of flora and fauna in the gardens and cultivated land destroyed by Vesuvius in A.D. 79, *in* Lockley, M. G., and Rice, A., eds., Volcanism and fossil biotas: Boulder, Colorado, Geological Society of America, Special Paper 244.

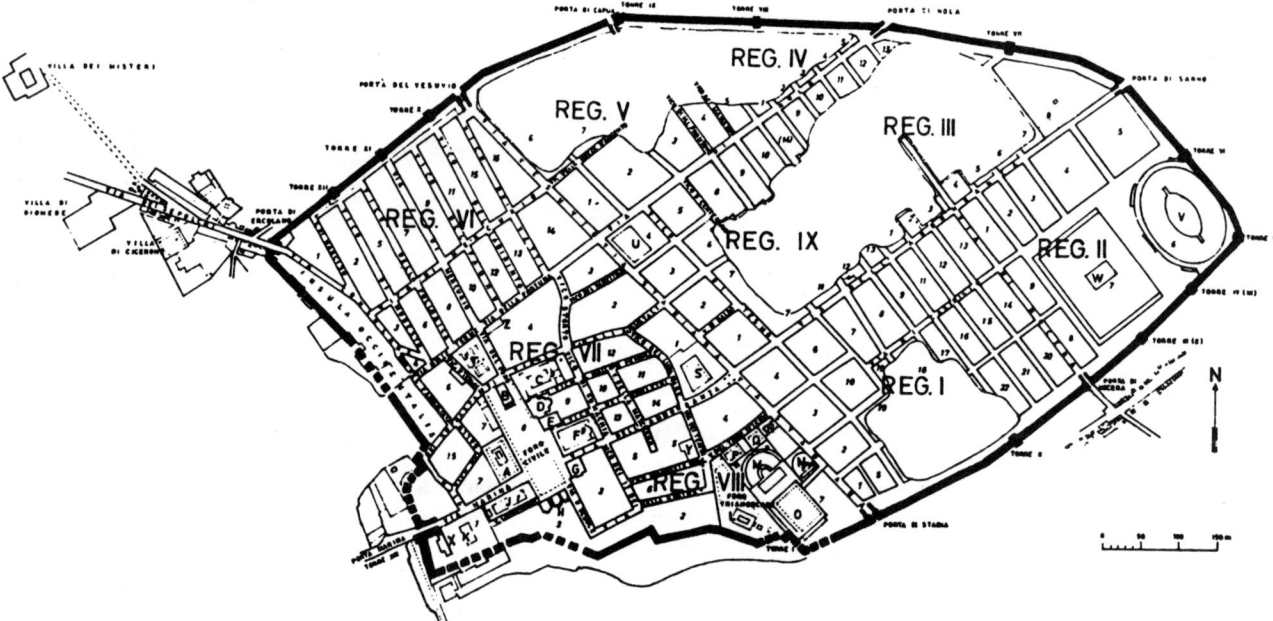

Figure 1. Plan of Pompeii. When Giuseppe Fiorelli became director of the excavations in 1860 he divided the city into 9 regions. Each region was subdivided into numbered *insulae*, or city blocks, and each entrance in each block was assigned a number. Thus, each door has an address of three numbers. The unexcavated area (about one-fourth of the city) is blank.

accounts of Pliny the Younger in chronologically reconstructing events of the eruption.

Gardens had a prominent place in the layout of Pompeii and Herculaneum, so we were hopeful that excavation would yield evidence of flora and fauna. There was at least one garden in almost every house; some had three or four. Luxurious villas had even more. The Pompeian built his garden in the interior of his house instead of surrounding his house with a garden as we do. Some of the older houses, however, had a rear garden planted with vegetables, fruit trees, and often a few vines. I have found that gardens played a hitherto unexpected role in restaurants, hotels, shops, and schools. There were also gardens connected with public buildings. More surprising was my discovery within the city walls of large commercial vineyards and orchards, as well as commercial flower and vegetable gardens (Jashemski, 1979). The discovery that valuable land within the city was so planted is of considerable importance in the study of land use and city planning.

The identification of all the planted areas, large and small, led me to make a comprehensive study of land use at Pompeii, for it is only as I have slowly identified and studied each individual garden that I have come to an appreciation of the layout of the city as whole. Pompeii is of unique importance in this respect, for scholars can do little more than speculate as to the probable uses of land in other ancient cities. After I had located every garden and planted area, it was possible to compute the percentage of space devoted to gardens, as well as to other purposes, and to note the way in which each function was distributed in the city. A summary of land use at Pompeii shows that buildings occupied 64.7 percent of the excavated area, streets and fora 17.7 percent, and gardens and cultivated areas another 17.7 percent. Of the latter, gardens attached to houses occupied 5.4 percent; large food-producing areas occupied 9.7 percent; gardens attached to various public buildings and businesses account for another 2.6 percent. More than one-third of the city was open space, perhaps even more if the entire city were to be excavated, for there are indications that a greater percentage of open space lies in the area still to be uncovered.

EXCAVATION TECHNIQUES: THE RETRIEVAL OF EVIDENCE OF FLORA AND FAUNA

The careful excavation of an ancient garden provides the only definitive evidence of its actual appearance and use in antiquity. The first step in the excavation of a planted area is the removal of the lapilli until the level of the soil in A.D. 79 is reached. At this point the cavities that were left by the decay of the ancient roots and into which the lapilli slowly trickled are clearly visible. Using special long-handled tools with tiny spoons at the end, or using long-handled tweezers, we carefully remove all the lapilli from the cavities. We fill the cavities with cement, which is allowed to harden for three or more days. When the soil around the cast is removed, the shape of the ancient root is revealed. The next step is the identification of the casts. Both the shape and size of the roots, and the way in which they are arranged in a planting pattern are significant.

The contour of the soil, perfectly preserved by the covering of lapilli, and the methods of watering also give valuable plant-identification clues. No new soil contour was discovered that puzzled us for long; one of our workers would always recognize it as similar to a contour in his garden, or in the garden of a relative or neighbor, which we would promptly examine. Great is the continuity of agricultural practices in the shadow of Vesuvius.

A minute examination of the surface of the soil sometimes yields carbonized or partially carbonized fruits, nuts, and vegetables (Jashemski, 1979). Carbonized fruits, nuts, and vegetables had previously been found in shops and cupboards (Meyer, 1980), but those that we have excavated are of special importance because they show the actual conditions of cultivation. Partially carbonized root material is sometimes found in root cavities, and occasionally we have found partially preserved twigs and even branches of trees in the upper lapilli. Once, when the cellular structure had been almost completely destroyed, a sample was treated with a combination of chromic and nitric acids by Dr. Francis M. Hueber, Curator of Paleobotany at the Smithsonian Institution; a microscopic examination of the preserved cells made it possible for him to identify the olive wood. The identification of ancient pollen adds further information about the flora (Jashemski, 1989).

Our excavations have also yielded information about the fauna of the area. Bones sometimes identify animals found in the garden. We have found the bones of at least one dog in every large garden that we have excavated, but the bones of only one cat. Bones are also often the refuse of meals eaten in the garden; sometimes they are evidence of sacrifices made to the gods. Shells add further information about the gardens.

Paintings found in the garden are an important source of information about the plants, animals, and even the fish of the area. One of the most striking aspects of many gardens is a picture painted on one or more garden walls to create the illusion of a much larger garden. In a very few houses, the walls of a room were painted to imitate a garden. In these paintings, often behind a painted fence that separated the actual garden from the painted one, there were trees, fountains, statues, and many birds, for which there was no room in the actual garden. I have been aided in the identification of the plants in these paintings by Dr. Frederick G. Meyer, research botanist in charge of the herbarium at the U.S. National Arboretum, Washington, D.C. He has botanized the Vesuvian area and found that the plants depicted in the paintings are those still found in the area today. These paintings, which unfortunately deteriorate when excavated and exposed to weather, are especially important because they picture the ancient flora; they form a unique chapter in the history of botany and cultivated plants (Fig. 2). The birds in the garden paintings are even more carefully painted than the flowers and trees. The birds have been identified by Dr. George Watson, Curator of Birds and Chairman of the Department of Zoology at the Smithsonian Institution, Washington, D.C., now retired, a specialist in Mediterranean birds. He has found that the birds depicted are typical of this area. A few garden paintings even include pools stocked

Figure 2. Detail of garden painting in the House of the Wedding of Alexander at Pompeii. A jay (*Garrulus glandarius* L.) is about to alight on a strawberry tree (*Arbutus unedo* L.) laden with fruit (left). To the right is an oleander (*Nerium oleander* L.) with pink flowers. Photo: Fotografia Foglia.

with carefully painted fish. These have been identified by Professor Eugenie Clark, Department of Zoology, at the University of Maryland, College Park. The fish are those of the area.

It seems very natural and charming to see the apparent size of a modest garden expanded through a garden painting. But, if the owner had greater aspirations, he might suggest that the painter include in his garden decorations not only fountains, birds, and plants, but also lakes and streams set in a mountainous landscape through which wild animals roamed in profusion. Great estates with large enclosures filled with wild animals were owned by many wealthy Romans in Italy at the time of the eruption of Vesuvius. What vast personal wealth made possible for a citizen of the Italian countryside could be suggested through

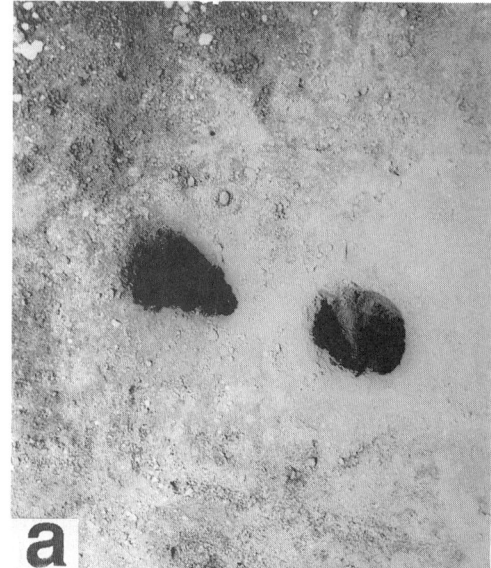

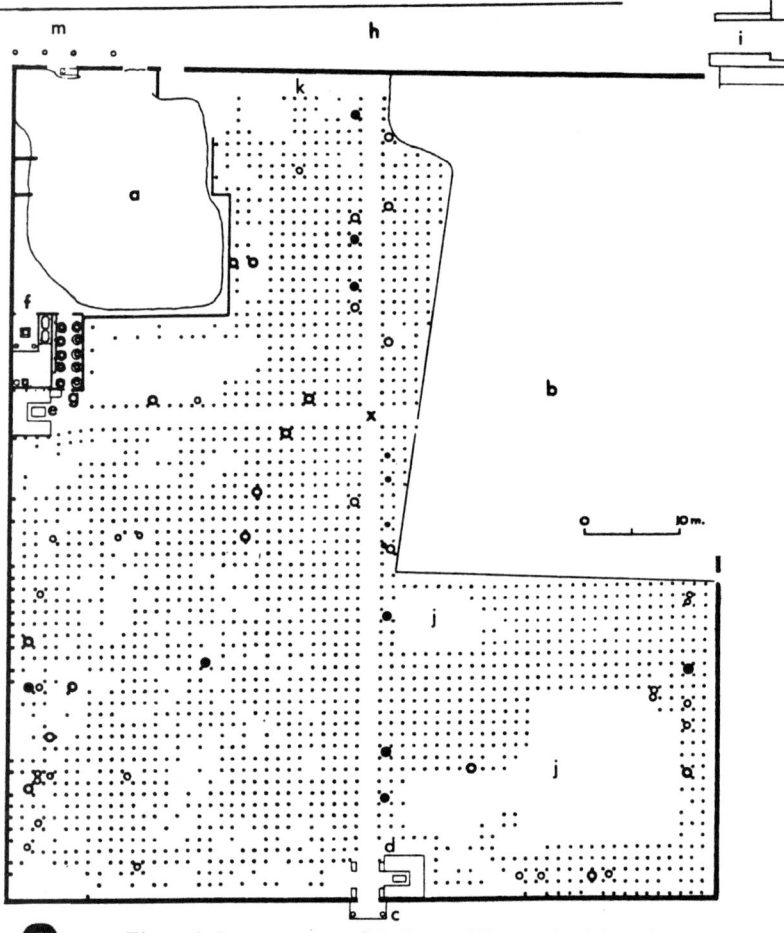

Figure 3. Large commercial vineyard II.v: a. Cavities left by stake and vine root. Photo: Stanley Jashemski. b. Cast of stake (left) and vine root with its side root reaching behind stake. Photo: Stanley Jashemski. c. Plan: (c) entrance, (d,e) triclinia, (x) intersection of paths; small dots indicate grapevine roots; circles, tree roots. Plan: Soprintendenza alle Antichità della Campania-Napoli; garden details: Stanley Jashemski.

the illusion of a painter's brush for the modest inhabitant of a town such as Pompeii or Herculaneum. The animals in these paintings have been identified by Dr. Henry Setzer, Curator of Mammals and Director of the African Mammal Project at the Smithsonian Institution, now retired. The animals are those that would have been known by the inhabitants of this area. (For a description of all the garden paintings in the Vesuvian sites, plus those found elsewhere in the Roman Empire, and the identification of the flora and fauna found in them, see Jashemski, 1989. This volume also contains descriptions of all the excavated gardens in the Vesuvian sites and tables of all spores and pollens found, as well as lists of all plant remains, mammal bones, mollusks, and echinoderms found.)

The enthusiastic cooperation of many scientists has made the study of the uniquely preserved gardens an exciting project that has crossed the boundaries of many disciplines and resulted in the bringing together of much specialized information. Some of the data accumulated cannot be attributed to a specific garden, but it is important for the entire area. A volume now in preparation, *The Natural History of Pompeii and the Other Vesuvian Sites,* will include chapters by 22 scientists who have worked with me, summarizing current evidence about the Vesuvian area in their particular discipline. The volume will include a detailed cooperative discussion of the causes and extent of the carbonization and preservation of biological materials at the various Vesuvian sites.

This chapter briefly reports the excavation of 6 very different types of sites, three at Pompeii and three at nearby villas, and discusses specific examples of the various types of evidence of flora and fauna found at these sites.

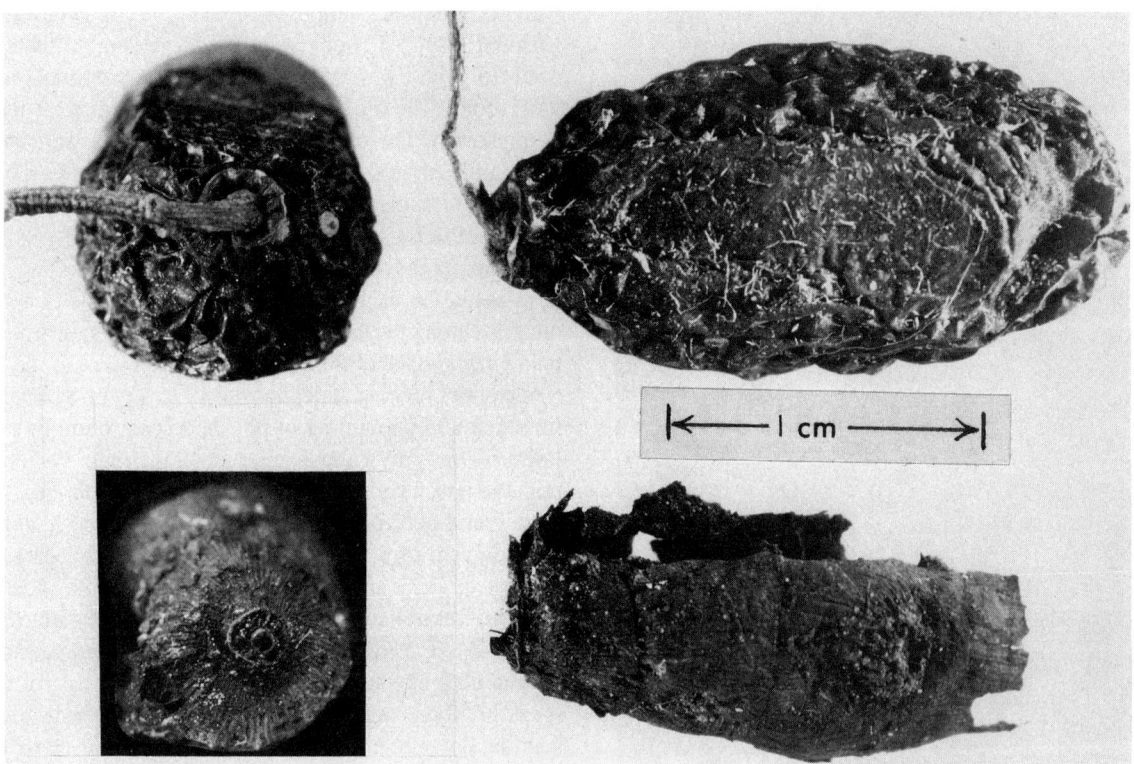

Figure 4. Olive from the U.S. National Arboretum Herbarium (top), and specimen found in vineyard (bottom). Photo: U.S. Department of Agriculture.

A LARGE COMMERCIAL VINEYARD AT POMPEII

For more than 200 years the large *insula* to the north of the amphitheater at Pompeii (Region II, *insula* 5) was known as the *Foro Boario* or cattle market, a name given to it by the early excavators in 1755, who had uncovered a small portion of it. Only subsoil excavation could determine if this valuable property had been planted. This I began in 1966. I discovered that the excavators who had uncovered a good part of the *insula* during the 1950s had removed the volcanic fill down to the A.D. 79 level, and later covered the excavated area with backfill. But I found that even in areas that have been previously excavated it is possible to find root cavities if modern roots have not been too destructive. The backfill was deep and compacted, making excavation exceedingly difficult and results discouraging, but we persisted and eventually found many grapevine root cavities.

The area had been too disturbed, however, for us to be certain whether the second cavity found at many locations was that of a stake or another root. The local growers were sure that the vines had been staked in antiquity just as they are today. Fruit experts at the University of Naples, on the other hand, believed that the vines had been pruned low, and left unstaked, and that the second cavity was also that of a vine root.

The importance of this vineyard made it desirable to excavate part of the portion still covered by the hill of original volcanic fill, and to examine the cavities in an undamaged area. We found that one cavity was always that of a vine root, easily identified by its shape and small lateral roots (Fig. 3a, b). The second cavity was always that of a stake. The local growers were not surprised. They knew that was what we would find! By the end of the 1970 season, we had found a total of 2,014 vine-root cavities and 2,014 stake cavities. The vines were also exactly four Roman feet (30.5 cm) apart (Fig. 3c). Around each root were three or four depressions for holding water. Two well-preserved examples of layering, to propagate new vines, were found.

We also discovered two intersecting paths dividing the vineyard, just as the ancient agricultural writers recommended. The cavities on each side of the path were perfectly preserved. The larger cavities were left by posts. The smaller cavities on each side of the post were left by vine roots. The posts had supported an arbored passageway similar to those found in vineyards in the Pompeian area today. The posts were probably chestnut, recommended in antiquity because of its obstinate durability and still preferred today. The ancient writers also recommended planting both the willow and poplar to furnish withes for tying the vines. Today the Pompeian vintner still ties his vines with the poplar and the willow. Fifty-eight trees were found in the vineyard. Two partially carbonized olives (*Olea europaea* L.) were found near one tree-root cavity (Fig. 4).

Fifty bones and 2 teeth were found in the vineyard (see Table 1 and Fig. 5). These were identified by Dr. Henry Setzer (Jashemski, 1979, p. 217). Eleven of the bones had cleaver

Figure 5. Cow bones split for marrow, tool marks visible. Photo: Stanley Jashemski.

marks, evidence that the bones had been split for marrow, which was considered an ancient delicacy. The bones were the remains of meals served at the masonry triclinia, or couches (made comfortable with pillows), where the ancient Pompeians reclined to eat. The discovery of these bones finally explained why the first excavators, who uncovered the area near the triclinium at the entrance of the vineyard, mistook this area for a cattle market.

A MARKET-GARDEN ORCHARD AT POMPEII

This once fine house (Region I, *insula* 15), known as the House of the Ship *Europa,* takes its name from the large graffito of a ship labeled *Europa,* found on the north wall of the peristyle when it was excavated in 1957. The house had obviously been converted to some kind of commercial use by the time of the eruption. The large, open, split-level area at the rear had been badly damaged at the time of the original excavation, but even so, when we excavated this area, we found two vegetable plots with distinct furrows in the lower garden, and a total of 416 root cavities in the entire garden. The many single cavities in the lower garden, regularly spaced 4½ Roman feet apart, were those of young vines (Fig. 6). They had distinct depressions for holding water on each side, similar to those in the large vineyard. The 31 irregularly spaced root cavities in the vegetable plots were undoubtedly those of small fruit trees, such as are found in vegetable plots today. The root cavities of the vegetables were too small to be preserved. The other cavities in the garden were the roots of trees, and perhaps some very old grapevine roots. The random way in which the trees were planted and their various sizes are duplicated in orchards in the area today.

Among the important discoveries in this garden were the 28 terra-cotta pots found embedded in the soil at varying depths, at intervals along the four walls. Eight of the pots were only slightly below the surface of the ground. Fourteen were at the bottom of root cavities ranging in depth from 17 to 42 cm. This was the first time that a large number of pots had been found in a garden at Pompeii, or to my knowledge at any Italian site. With one exception, the pots have 1 hole in the bottom and 3 on the sides. I have since found pots, or fragments of such pots, in several gardens in the Vesuvian sites. The question arises as to how these pots had been used. There are few references to pots by the ancient writers. The agricultural manual written by the farmer-statesman Cato (234–149 B.C.) and Pliny the Elder's encyclopedic *Naturalis Historia* contain helpful information about ancient gardening. They speak briefly of using pots to make air layers, or layers made by bending down branches planted in a pot into the earth, if they are to be carried a considerable distance (Cato 52.133; Pliny 17. 97, 98). Pliny (12.16) also reports that "various countries tried to acclimatize the citron, importing it in earthenware pots with breathing holes for the roots." This is a good description of our Pompeian pots. Is it possible that in this garden there were exotic trees, such as the citron, that were imported in pots? At the time that Pompeii was destroyed there was great interest in introducing new fruit and nut trees into Italy. Perhaps the ship *Europa* did reflect the commercial activity of the owner. Or could lemon trees perhaps have been planted in the pots? Today at Pompeii, lemon trees are air layered in pots, or their modern equivalents. But perhaps other trees were layered in these pots.

We found other important evidence that helps to identify what was grown in this garden. A considerable number of carbonized fruits, nuts, and vegetables were found preserved where they were grown. These were identified by Dr. Frederick G. Meyer. There were pieces of filbert shells (*Corylus avellana* L.), a piece of almond shell (*Prunus dulcis* (Mill.) D. A. Webb), perfectly preserved grapes (*Vitis vinifera* L.), caramelized because of their high sugar content, and even grape seeds, as well as a piece of carbonized fig (*Ficus carica* L.). The large number of broad beans, or horse beans (*Vicia fava* L. var. *minor* (Peterm. and Harz) Beck), found in various places in this garden, strongly suggests intercultivation among the vines and trees, which is practiced extensively in the Pompeii area today. With tiny tweezers, entomologists at the Smithsonian Institution extracted the hind leg of a bruchid, or strawberry weevil, from a hole in one of the ancient beans. From another they extracted a large part of a bruchid.

Twenty-nine bones and 20 teeth, representing 5 animal species, were found during our excavation of this garden. They were identified by Dr. Henry Setzer as follows: dog, 7 bones; pig, 8 bones and 8 teeth; cow, 4 bones; sheep, 3 bones and 12 teeth; sheep or goat, 3 bones; chicken (*Galus gallus* L.) 1 bone; and 3 bones too fragmentary to identify. We have no way of knowing how many bones were removed when most of the lapilli was trucked away at the time of the original excavation. The dog was probably a watch dog or pet. The two femurs, which are from the same dog, indicate that the dog was about 45 cm high at the shoulder. The other bones may be those of domestic animals that were trapped in the garden during the eruption. It is possible that animals may have been kept in the building in the southeast corner of the *insula*, for it was not uncommon to find animals within the city. There were no tool marks on the bones to indicate that they had been split for marrow. Nor was there a triclinium in this garden. But it appears to have been a usual practice to discard bones in the garden after a meal.

A SMALL FORMAL GARDEN AT POMPEII

Most of the houses at Pompeii had a garden in the middle of the house. It might be a simple courtyard or a peristyle garden, with a portico on one or more sides. Unfortunately most of these were excavated before the days of scientific garden excavation, and we know little about their actual plantings. The planting pattern in the small house (Region I, *insula* 12, entrance 11) directly across the street from the House of the Ship *Europa* was very different from anything that we had yet found. The little garden was enclosed by a portico on three sides, with a huge but badly preserved animal painting on the rear wall. The house had been previously excavated and most of the lapilli had been removed from the garden, but fortunately the area was free from destructive weeds, and modern roots had not destroyed the evidence of ancient ones. The root cavities that we found in 1975 were those of small shrubs, ranging in size from 2 to 4 cm in diameter at ground level. They were planted in a formal design (Fig. 7a). In the center of the garden was a small statue base framed by the plantings, which would have been evergreen, probably box.

THE *VILLA RUSTICA* IN THE *LOCALITÀ* VILLA REGINA AT BOSCOREALE

At the time of the eruption the countryside was dotted with an extraordinarily large number of villas. Unfortunately, of those discovered, few are still visible. Many were found on private property and only partially excavated, for the owners were anxious to fill in the temporary excavations and return the valuable land to cultivation. These villas differ greatly in size, luxuriousness, and function. Most appear to be real agricultural estates (*villae rusticae*).

Recently, during construction of a new apartment complex at Boscoreale, on the lower slopes of Vesuvius just 1 km north of Pompeii, 80 large cement pillars were put in the ground and evidence was found of an ancient rustic villa. Fortunately, the villa and the land immediately surrounding it were declared a permanent archaeological zone. The villa building was excavated in 1977–1980 by Dr. Stefano De Caro, Director of the Excavations at Pompeii. This site is an unusually important one, for it is the first time in the entire Vesuvian area that there has been an

Figure 6. Market-Garden Orchard I.xv. Balloon photo shows soil contours of vineyard and two vegetable plots in lower garden. Photo: Julian Whittlesey Foundation.

Figure 7. a. Formal planting of clipped hedge in peristyle garden. Drawing: Luc Herbots. b. Country road in front of villa rustica at Boscoreale showing casts of tree roots along the road. Photo: Francis Hueber.

opportunity to excavate farmland attached to a villa. I excavated this land during the summers of 1980, 1982, and 1983. We found that most of the land had been planted in a vineyard. The vines, which were informally staked, were supported by both stakes and trees, as is true in many vineyards in the area today. There were 34 tree-root cavities, most of which were in the vineyard. The partially carbonized olives and almonds found in the vineyard help to identify some of the trees. There were also many grape seeds and grape stems.

The villa building contained two rooms in which the grapes were pressed. After the juice was extracted it was fermented in the 18 dolia embedded in the peristyle courtyard.

To the right of the main entrance of the villa was a small vegetable garden, with a cistern in the middle to provide the necessary water. Here, no doubt, were raised the choice cabbages and onions for which the Pompeii area was famous in antiquity. The garden was divided into small plots separated by irrigation channels, which also served as paths.

A roadway led from the main entrance to the country road that passed in front of the villa. Along the edge of the country road, on the villa side, there was a row of huge trees, quite unlike the small fruit and nut trees in the vineyard (Fig. 7b). Analyses of the woody material found in the root cavities made it possible for Dr. Francis M. Hueber, who identified our wood samples, to identify tree 29 (50 × 55 cm at ground level) on the left of the roadway to the villa, and also tree 30 on the right (56 × 80 cm at ground level) as umbrella pines (*Pinus pinea* L.). The two other large trees, 27 and 28, were definitely angiosperms (broadleaved deciduous trees), but the woody material analyzed so far has been too damaged to make it possible to identify the trees as to species. The shape and size of the root cavities suggest that they were plane trees (*Platanus orientalis* L.).

Not surprisingly, Dr. Eberhardt Grüger, professor of botany and palynology at the University of Göttingen, who analyzed our soil samples for pollen, found a large amount of *Pinus* pollen near the umbrella pines. Among the pollens found were *Vitis* (vine), *Olea* type, *Triticum* (wheat), and *Arbutus unedo* (strawberry tree). However, analysis of the soil samples from Boscoreale has only begun.

I wondered if the contents of the amphoras or other containers found in the kitchen might reflect crops raised on the villa. To determine the contents, scrapings were made. After preparation, each sample was viewed in the scanning electron microscope, and an energy-dispersive spectrometer x-ray was performed by Dr. M. E. Taylor, head of the electron microscope facility, and Dr. Franz Kasler, professor of chemistry, both of the University of Maryland. This showed that the contents included wine, oil, and garum. Wine was produced at the villa. Olives were raised, but no press has yet been found. Garum, a popular fish sauce, would certainly be in any well-stocked kitchen.

Cement poured into a cavity in the volcanic ash preserved the appearance of a pig raised on the villa. Among the bones found in the vineyard were those of pig, sheep or goat, cow, a toad or frog, and a dog. The bones of a coot (*Fulica atra* L.), a water bird, were found in the trash pile (which included debris from meals) located around the tree near the main entrance of the villa.

Debris from meals and other refuse was also found at the edge of the neighboring property on the south side of the country road. The bones of a rail (*Rallus aquaticus* L.), another water bird, were found here. Only the bill of a chaffinch (*Fringilla coelebs* L.) was found, but small song birds were considered a table delicacy. The bird bones were identified by Dr. Storrs L. Olson, Curator, Department of Vertebrate Zoology at the Smithsonian Institution. Bones of a dormouse (*Muscardinus avellanarius* L.), also a table delicacy, were also found here. More difficult to explain are the bones of a little ermine or weasel (*Mustela erminea* L.) and the bones of a martin (*Martes martes* L.). There was also the partial skeleton of a small rodent, the pine vole

(*Pitymys savii* de Selys Longchamps), and the almost complete skeleton of a rat snake. This adjacent property was also planted in vines.

THE VILLA OF L. CRASSUS TERTIUS AT OPLONTIS

No garden has yet been found in this *villa rustica*, found by chance during construction of a school, at a site near the sea identified as ancient Oplontis (modern Torre Annunziata), 5 km from Pompeii. In one of the rooms opening off the peristyle, however, there were several cubic meters of carbonized plant material that apparently had fallen from the upper story where it was stored. Although no farmland connected with the villa has been found, this totally unexpected find gives us an opportunity to examine specimens of vegetation grown on land connected with the villa.

This carbonized material is being separated, studied, and photographed by Dr. Massimo Ricciardi, professor of botany at the Institute of Botany of the University of Naples Agricultural College at Portici, and Dr. Francis Hueber. Professor Ricciardi has identified, thus far, 128 taxonomic entities in this hay (Ricciardi and Aprile, 1978), adding 81 species, 37 genera, and one family to the list of 408 plants previously thought to have been known to the Romans in the first century after Christ (Fig. 8). The large quantity of grapevine leaves, tendrils, and small branches, some of which clearly showed the cut where they had been pruned, makes Professor Ricciardi believe that this hay was collected in a vineyard. *Prunus,* olive, and oak leaves were also present. The majority of the plants identified were legumes or grasses such as would have been found growing under the vines. These would have made good hay. There were also the ever-present weeds and wild flowers, including various sorrels, chickweed, mustard, mints, buttercups, wild geraniums, mallow, little

TABLE 1. ANIMAL REMAINS FOUND IN VINEYARD*§

Field Number	Bone	Tool Marks	Probable Tool Marks	Field Number	Bone	Tool Marks	Probable Tool Marks
	Bos taurus L. (domestic cow)				**Sus scrofa L. (wild boar or domestic pig)**		
1	Metacarpal			12	Scapula		X
3	Metatarsal, distal end	X		36	Scapula (fragment)		
9	Os coxae			37	Lower mandible (fragment)		
10	Metatarsal	X		38	Scapula (fragment)		
11	Metatarsal	X		44	Tarsus center (fragment)		
14	Metatarsal, proximal end			50	Piece of mandible		
15	Metatarsal, distal end		X		**Canis familiaris L. (domestic dog)**		
29	Molar			6	Os coxae		
30	Molar chip (from No. 29)			7	Radius		
31	Femur, distal end	X		17	Radius		
34	Calcaneum	X			**Ovis aries L. (domestic sheep)**		
35	Second phalanx			33	Femur, distal end		
45	Metatarsal						
47	Metatarsal (fragment)	X			**Ovis aries or Capra Hirca L. (domestic goat)**		
2	Scapula*			51	Metapodial		
	Equus caballus L. (domestic horse)				**Felis catus (domestic cat)**		
5	Ulna	X		52	Lower left mandible		
19†	Tibia, distal end						
20	Tibia, center section near proximal end	X					
21	Radius						
23†	Tibia, proximal end						
32	Humerus, distal end (fragment)						
25	Rib*						
	Bos taurus or Equus caballus						
18	Vertebrae	X					
22	Calcaneum						
43	Metatarsal or metacarpal (fragment)						
46	Rib		X				

*Identification of species not entirely certain.
†Part of the same bone.
§After Jashemski, 1979, Table A (14 bones were indeterminable; of these, 2 had tool marks).

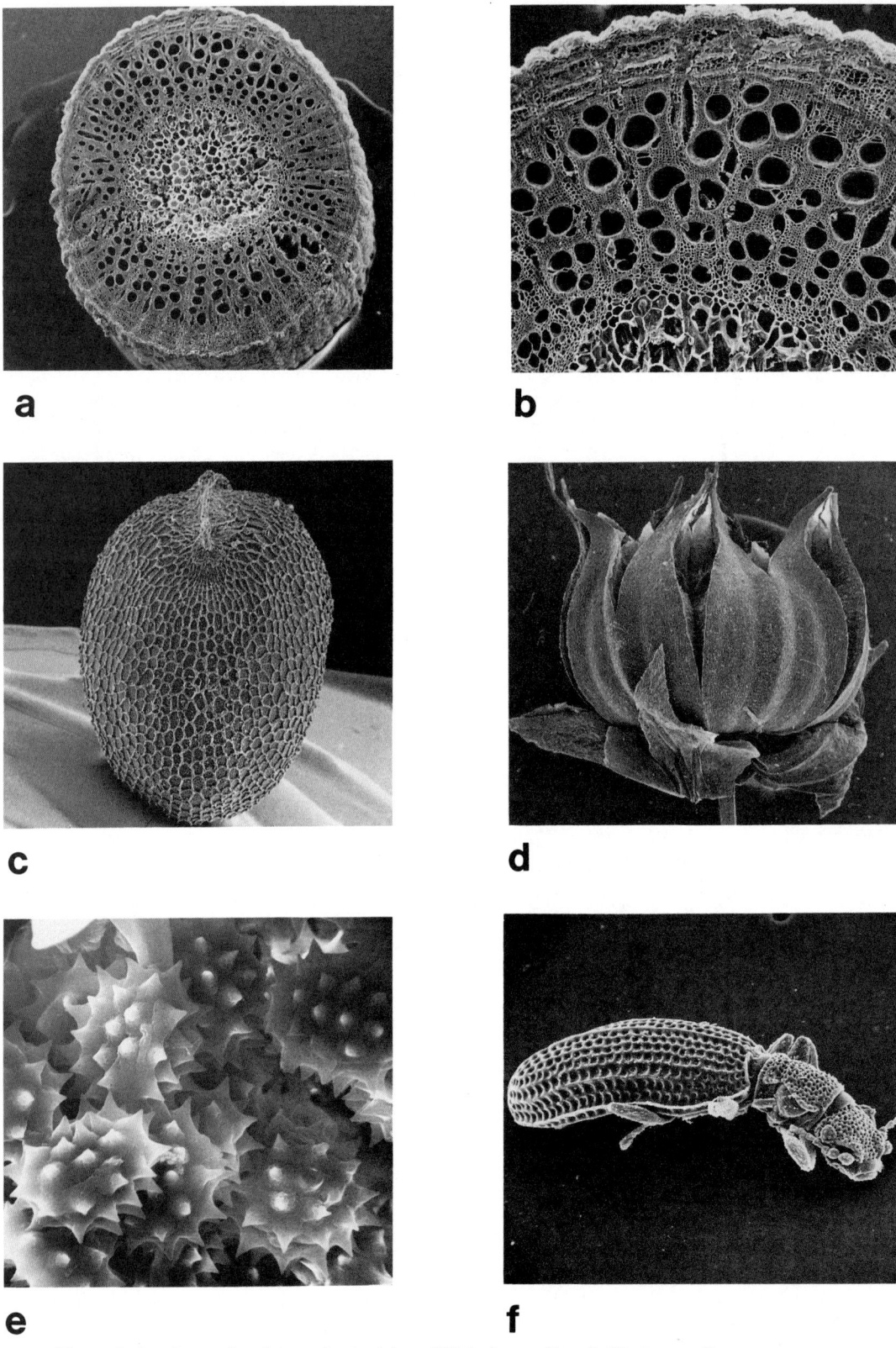

Figure 8. Specimens found in carbonized hay. SEM photos: Francis Hueber. a. Grape-stem cross section. b. Detail of grape-stem cross section. c. Geranium seed (*Geranium rotundifolium* L.). d. Flax capsule. e. Pollen grains (daisy family). f. Minute brown scavenger beetle (*Microgramme ruficollis* Marsham, family Lathridiidae). Identification by John Kingsolver, U.S. Department of Agriculture Research Entomologist at the Smithsonian Institution.

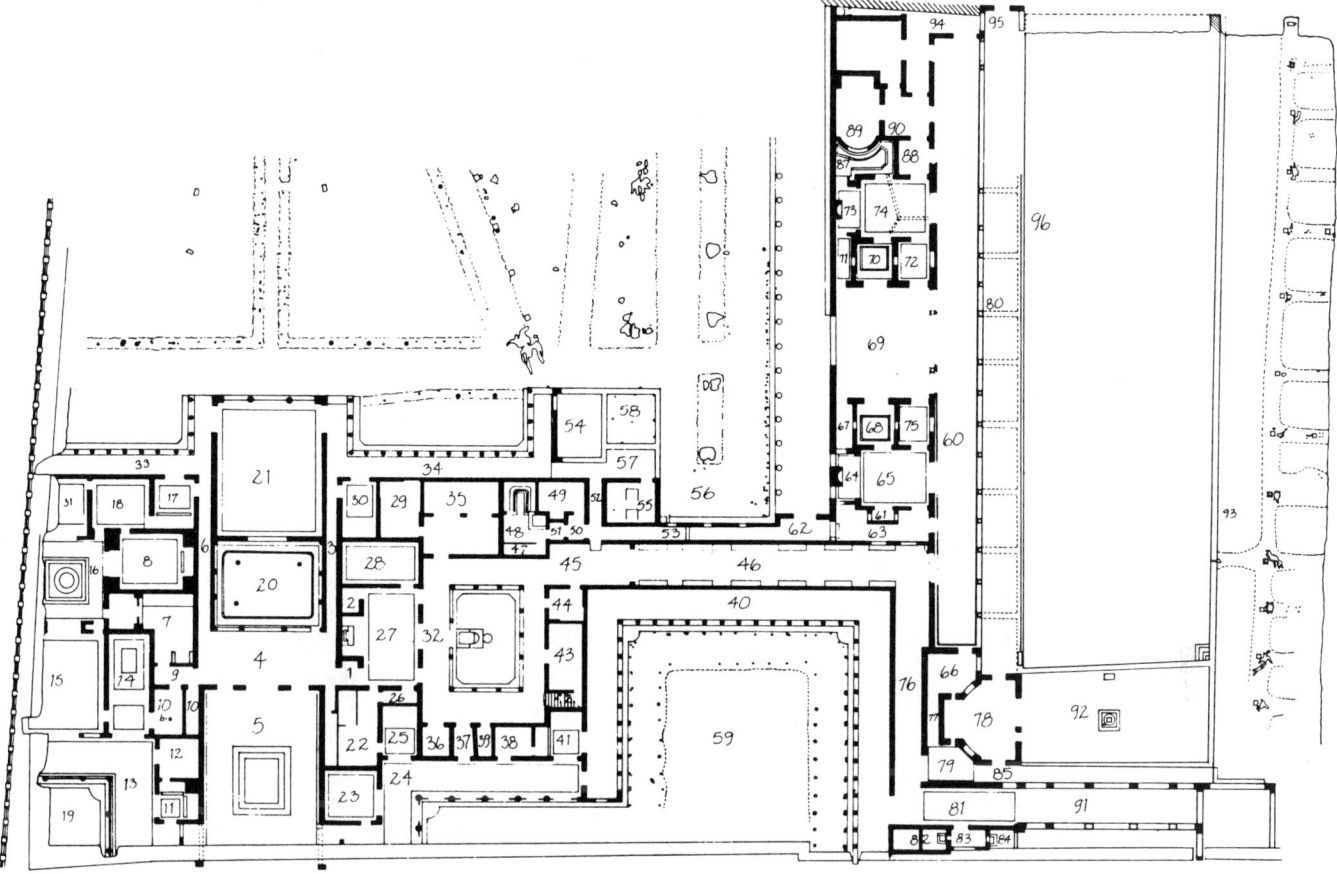

Figure 9. Villa of Poppaea at Oplontis. Plan: Soprintendenza alle Antichità della Campania-Napoli; garden details by Stanley Jashemski.

red poppies, chrysanthemums, and of course, brambles and bracken fern. There were also violet flowers (*Viola arvensis* Murray). The abundance of flax (*Linum usitatissimum* L.) in the hay indicates that this crop was grown nearby and had escaped into the vineyard.

The villa has yielded other significant evidence regarding the crops grown. The imprints of filberts that had been gathered and stored in the villa were found perfectly preserved, even though the nuts had completely disintegrated. In 1984 more than a ton of pomegranates (*Punica granatum* L.) that had been carefully stored between layers of straw were found in the villa.

THE LUXURIOUS VILLA OF POPPAEA AT OPLONTIS

This luxurious villa, very different from the two rustic villas described above, is believed to have belonged to Poppaea, the wife of the Emperor Nero. Excavation by the Superintendency of Naples began in 1964. Thus far, the villa has 13 remarkable gardens, which I have excavated (Fig. 9). For the first time, villa gardens could be scientifically excavated, using the most recent archaeological techniques. This villa is also important because it is the only one in which we have any archaeological evidence for the great exterior gardens onto which such villas opened.

The layout of the gardens in this villa is intimately related to the villa architecture. The original villa, built during the middle of the first century B.C., was compact in plan, with a large atrium as a nucleus and several interior gardens. The rooms of the east wing, built during the empire (A.D. 50 to 70), looked out on an Olympic-size swimming pool (60 × 17 m) and two handsome gardens, one at the south end and a great sculpture garden to the east of the pool.

In order to appreciate the relation of this sculpture garden to the total planning of the villa, let us briefly look at the great park-like garden (56) at the rear of the villa. The architectural layout of this garden reflects the plan of the villa itself. From the villa entrance, the perspective through the atrium (5) continued through an enclosed garden (20) and the grand room (21), with a monumental entrance at the rear to a landscaped pathway on the central axis of the villa. To the east of the central passageway was a diagonal one, obviously balanced by a similar one on the opposite side, which today is mostly under the modern street. The two diagonal paths probably met at the central passageway, at some point under the still-unexcavated lapilli to the north. At the east

Figure 10. Sculpture garden towards south, showing casts of tree roots. Photo: Francis Hueber.

edge of the garden were two parallel passageways separated by a row of huge trees, identified by their root structures as probably plane trees at least 100 years old.

The large rooms (65, 69, 74) in the east wing were all sited to take advantage of the landscaping. The large, elaborately decorated, central room (69), comparable to the one (21) at the rear of the villa, had an enormous window on the west, which looked out on the park-like garden at the rear of the villa. On the east, this room was completely open to the view of the large sculpture garden (93) beyond the swimming pool.

In this large sculpture garden an avenue of 13 trees (thus far) has been found (Fig. 10). In front of each tree was a statue base; six of the life-size statues that stood on them have been found. At the south was the head of Hercules on the marble pillar in front of cavity III. Next was an ephebe in front of root cavity IV, and next a large Nike. Balancing these statues, in corresponding positions along the north end of the pool, was another Nike in front of root cavity X; then a statue of Artemis or an Amazon in front of cavity XI; next a head of Hercules on the marble pillar in front of root cavity XII.

Both the shape of the root cavities behind the statue bases and later the casts were carefully studied by Dr. Carlo Fideghelli, from the Ministero dell' Agricoltura at Rome, a specialist in modern roots who always examines our root cavities and casts. Counting from the south, root cavities I, II, IV, V, X, and XI were all those of large trees. The fragments of branches found behind base I were too deteriorated to identify the tree as to species, but it was definitely an angiosperm. The size and shape of the root cavities led Dr. Fideghelli to identify these trees as probably plane trees. They remind us of the avenue of older plane trees in the large garden at the rear of the villa. Only root cavity XIII seems to depart from the symmetry of the plantings. Instead of a plane tree in this location, which symmetry would suggest, this tree left a single root cavity that was the same diameter (ca. 25 cm) down to a depth of 78 cm, at which point it divided into 5 large roots, the longest 57 cm. The root cavity has the appearance of a cypress and suggests that at the north end of the pool the garden may have made a transition to another planting pattern. It is interesting that Professor Grüger found cypress pollen in the soil samples taken in this part of the garden. The smaller root cavity behind each of the pillars with the head of Hercules contained no root material and cannot be identified. The tiny roots in the beds behind the pillars may have been those of myrtles (*Myrtus communis* L.), of which we found much pollen.

The ornamental plantings behind statue bases VI through IX, which were opposite the large room in the middle of the east wing, were carefully planned to provide a colorful and dramatic picture when viewed across the water. A scanning-electron-microscope (SEM) photograph of a tiny scrap of a branch of tree VI, extracted from a small cavity in the cut at the edge of the unexcavated lapilli, could be identified as oleander (Fig. 11). Root IX, which was similar in size and shape, also appears to have been an oleander. At a distance of approximately 3.4 m behind statue bases VI and IX we found a low base, and behind each of these were the plentiful remains of a small tree, which SEM photographs showed were also oleanders.

SEM photographs of the woody material from the tree behind statue base VIII indicated to Dr. Francis Hueber that this could be either a laurel (*Laurus nobilis* L.) or a lemon tree (*Citrus limon* (L.) Burm. f.), but when we excavated the root cast we found that this tree had been air layered in a pot. Laurels are easily rooted and are never started in this way. A broken amphora (the top half) had been used for a pot; the root grew out of the mouth of the amphora. Today, in this area, lemons are air layered, but old tin cans or plastic containers are the broken amphoras of the modern world. The root cavity behind statue base VII was similar in size and shape, and also appeared to be that of a lemon tree.

The complexity of this garden, which extends farther to the north, south, and east, will become clearer only with further excavation, but we are charmed by the portion already uncovered. Dramatically placed behind each of the two central statue bases was an exotic lemon tree, such as we find pictured in the garden paintings, and which were esteemed for the fragrance of their blossoms and the beauty of their fruit. Completing this picture, behind the statue base on each side of the lemon trees was a lovely oleander, with still more oleanders behind the small bases at the rear. The oleander would have been the pale pink one, ever present in the garden paintings, and so often pictured amid sculpture. This colorful scene of lemon trees and oleanders was framed on each side by three stately plane trees, which furnished shade to those strolling along the pool and cast shadows on the shining marble statues.

Slowly the Vesuvian gardens yield their secrets. The tragedy that overwhelmed these sites uniquely preserved an abundance of evidence that is richly increasing our knowledge of ancient Roman flora and fauna. The possibility of examining such evidence (which includes an abundance of pictorial material) together with the comments of the ancient writers and in the light of modern practices, which so strikingly continue those of antiquity, exists at no other site in the ancient world.

ACKNOWLEDGMENTS

Support for my excavations was provided by the National Endowment for the Humanities, Dumbarton Oaks, and the Soprintendenza Archeologica di Pompeii. My work was greatly facilitated by the generous cooperation and gracious hospitality of Dr. Giuseppina Cerulli Irelli, Superintendent of Pompeii; Dr. Stefano De Caro, Director of the Excavations at Pompeii; Ferdinando Balzano, assistant at Oplontis; and Vincenzo Matrone, assistant at Boscoreale. Nicola Sicignano was our able foreman.

I am also greatly indebted to the many scientists, in the United States and Europe, who through the years have cooper-

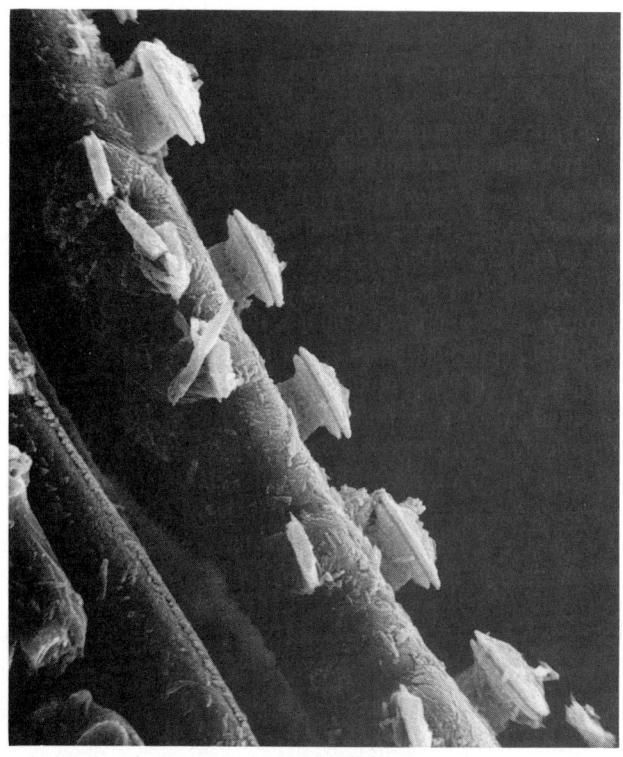

Figure 11. Vessel elements from branch of oleander found near statue base VI. Photo: Francis Hueber. Magnification ×2750.

ated with my work in the Vesuvian gardens. The specific contributions of 11 of these scientists are acknowledged in the discussions herein.

REFERENCES CITED

Cato, *De Agricultura,* 1954, Latin text with translation by Hooper, W. D., revised by Ash, H. B.: Cambridge, Massachusetts, Harvard University Press.

Jashemski, W. F., 1979, The gardens of Pompeii, Herculaneum, and the villas destroyed by Vesuvius, v. 1: New Rochelle, New York, Caratzas Brothers Publishers, 372 p.

——, 1989, The gardens of Pompeii, Herculaneum, and the villas destroyed by Vesuvius, v. 2: New Rochelle, New York, Aristide D. Caratzas, Publisher (in press).

Meyer, F. G., 1980, Carbonized food plants of Pompeii, Herculaneum, and the villa at Torre Annunziata: Economic Botany, v. 34, no. 4, p. 401–437.

Pliny the Elder, *Naturalis Historia,* 1938–1962, Latin text with translation by Rackham, H., and others, 10 volumes: Cambridge, Massachusetts, Harvard University Press.

Pliny the Younger, *Epistulae,* 1969, Latin text with translation by Radice, R., 2 volumes: Cambridge Massachusetts, Harvard University Press.

Ricciardi, M., and Aprile, G. G., 1978, Preliminary data on the floristic components of some carbonized plant remains found in the archaeological area of Oplontis: Annali della Facoltà di Scienza Agrarie dell'Università di Napoli in Portici, series 4, v. 12, p. 204–212.

Sigurdsson, H., Carey, S., Cornell, W., and Pescatore, T., 1985, The eruption of Vesuvius in A.D. 79: National Geographic Research, v. 1, part 3, p. 332–387.

Manuscript Accepted by the Society June 21, 1989

Typeset by WESType Publishing Services, Inc., Boulder, Colorado
Printed in U.S.A. by Malloy Lithographing, Inc., Ann Arbor, Michigan

JUL 2 4 1990

JUL 2 4 1990